Palgrave Studies on Norbert Elias

Series Editor
Tatiana Savoia Landini
Universidade Federal de Sao Paulo
Sao Paulo, Brazil

Despite growing, widespread appreciation for Norbert Elias's theoretical approach—often called figurational or processual sociology—there exist only a few, specialized publications on Eliasian social theory, and as of yet, no academic book series.

Palgrave Studies on Norbert Elias will therefore fill a significant gap in the market, appealing to figurationalists across disciplines: Elias's social theory is used not only in Sociology, but also Sports, Psychoanalysis/Psychology and Social Psychology, Education, Criminology, International Relations, History, Humanities (Arts, Music, and Cultural Studies), Political Science, and Public Health. Respecting the multi-disciplinary Eliasian tradition, the series is open to receiving contributions from academics outside of Sociology departments, so long as the research is grounded on Elias´s approach. Publications, which shall range from Palgrave Pivots to edited collections, can be expected to explore sports, habits and manners, criminology, violence, group relations, music and musicians, theory and methods, civilizing and decivilizing processes, involvement and detachment in social sciences, formation of the modern state, power relations, and the many dozens of other topics to which Eliasian theory has been applied.

More information about this series at
http://www.palgrave.com/gp/series/16153

Debbie Kasper

Beyond the Knowledge Crisis

A Synthesis Framework for Socio-Environmental Studies and Guide to Social Change

palgrave
macmillan

Debbie Kasper
Environmental Studies
Hiram College
Hiram, OH, USA

ISSN 2662-3102 ISSN 2662-3110 (electronic)
Palgrave Studies on Norbert Elias
ISBN 978-3-030-48372-2 ISBN 978-3-030-48370-8 (eBook)
https://doi.org/10.1007/978-3-030-48370-8

This Palgrave Macmillan imprint is published by the registered company Springer Nature Switzerland AG
The registered company address is: Gewerbestrasse 11, 6330 Cham, Switzerland

For Tim.
Thanks for helping me cover the ball.

Acknowledgments

In the spirit of interdependence, I'd like to formally acknowledge the countless others (widely dispersed across space and time) who have contributed to shaping my thinking and worldview. It is thanks to their labors that I have anything to say at all; any errors or shortcomings in what I say in these pages are entirely my own.

Surely, there are many influences of which I'm not consciously aware. Of those I *am* aware of, there are some I'd like to explicitly recognize here. Gratitude first goes out to all those who have served as teachers in my life, beginning with my family, continuing with friends, and including all those whom I've encountered through years of formal education. You have contributed to shaping my sense of self and the world more than I can say, or will probably ever know. Along those lines, Jim Benton and Alan Sica merit special mention. Whether you realize it or not, the lessons, advice, and casual comments you proffered to the younger me-in-progress pointed toward a path of discovery that continues to unfold in ways I could have never predicted.

Especially noteworthy was my initial encounter with Norbert Elias as a graduate student in one of Sica's classes. Trusting my sense that Elias' work offered critical insights which the world needs, especially now, has paid dividends in the form of increasing clarity about human social life and how to study it. As a bonus, it has also connected me with a small but mighty cadre of scholars devoted to sharing those insights with the world. I am extremely grateful to these fellow appreciators of Elias (spread around Europe and the rest of the globe as they are) for their brilliant scholarship and for making my academic world less lonely over here on this side of the Atlantic.

I would like to acknowledge the many students I've had the pleasure to know over two decades of teaching who have also served as my teachers in so many respects. I extend special appreciation to those who have read select chapters of this book and, in particular, to the curious and energetic crew in Systems Thinking and Social Change: Alexander Andrzejewski, Zack Fox, Julia Goetz, Rashid Jackson, Kasie Purpura, Henry Schwendler, Elijah Skaggs, and Kyle Workman. Your enthusiasm, insights, and feedback were crucial for giving me the confidence to keep moving forward with this project. Extra props go to Zack, who, with his sharp eye and unfaltering work ethic, was an enormous help in the final phase of completing the manuscript. Thanks also to my colleague Sarah Mabey, whose encouragement and camaraderie over the years have been invaluable and who so capably held down the fort as I worked to finish this book, and to Dean Judy Muyskens without whose support this project would have taken much, much longer.

Having worked on this book long before I was certain it would ever turn into anything, I want to express deep gratitude to those who saw something of value in it and shepherded me through the process of becoming a contributor to this series. This includes series editor Tatiana Landini, commissioning editor Mary Al-Sayed, anonymous reviewers who offered valuable feedback, and editorial assistant Madison Allums. Your impeccable professionalism and the gracious guidance you provided at each step along the way are profoundly appreciated.

Finally, I acknowledge those to whom I am most indebted: my husband, Tim, and children, Lucía, Theadora, and River. Not only have you offered endless encouragement (and many other forms of sustenance!), you have also put up with years of my out-loud brainstorming, incessant self-doubt, and spending far too much time at a computer rather than hanging out with you. I can only hope that this book offers some benefit to the world in ways that might begin to make up for it.

Praise for *Beyond the Knowledge Crisis*

"Although not *about* Norbert Elias, *Beyond the Knowledge Crisis* is very much in Eliasian spirit. Broad in sweep, Kasper's book examines the ecological crisis—*the* practical issue of politics and policy now and in the near future. The author's stress throughout is on the word *synthesis*, recognizing that a theoretical and empirical intellectual synthesis of history and the social and natural sciences is necessary to tackle the global problems of the age. This very synthesis is that which Elias himself was seeking to achieve, especially in his late writings."

—Stephen Mennell, *Professor Emeritus of Sociology, University College Dublin, Ireland*

"Debbie Kasper's *Beyond the Knowledge Crisis* is a vital work that arrives at a critical time. Scholars of Norbert Elias should be delighted with Kasper's explorations of his work and influence, but perhaps more importantly, all practitioners of integrative problem solving will find great insight and guidance here. Kasper provides a much-needed tonic during a time of enormous practical and existential challenges, when teachers and students alike need clear guidance to undertake integrative problem solving, especially in environmental studies. As an iconoclastic interdisciplinarian with a 30-year career learning and teaching, I can attest that it is rare to encounter a book that so skillfully and accurately depicts the world of professional practice that I have both inhabited and researched for so long. Kasper's work will help us navigate the complexities of setting and achieving truly interdisciplinary goals in the face of the great complexity that now demands them."

—Richard L. Wallace, *Professor of Environmental Studies and Marine Science, Chair of the Environmental Studies Department, Director of the Food Studies Program, and Co-Director of the Whittaker Environmental Research Station, Ursinus College, USA*

Introduction

One finger's-breadth at hand will mar
A world of light in heaven afar,
A mote eclipse a glorious star,
An eyelid hide the sky.

John Keble (1851:65)

The width of a finger, a speck of dust, an eyelid. The smallest of things can obscure our view of even the vastest panoramas. When I first encountered the above bit of verse, I was struck by the eloquence with which it captured this profound truth. I have experienced firsthand how a seemingly minor shift in perspective can lead to major differences of interpretation and understanding, and how the scales can fall from one's eyes in the blink of an a-ha moment.

Being somewhat of a collector of quotes, I marked the page and added the book to a stack of items containing others' words destined for my files. When I finally got around to typing it up, I was using an old keyboard on which the nubs of the home row anchor keys "F" and "J" had been worn away. Without knowing it, my hastily placed fingers were one key to the left of where they should have been. I was rapidly typing away when the odd-looking configuration of letters on the screen finally caught my attention. Looking more carefully, I saw that what I had produced was sheer nonsense.

Ibw dubfwela0vewSRG r gBS qukk nE
QIEKS ID KUFGR UB GWcwb Dem
NIRW WXKUOAW fkieuiya arEM
B wtwkus gusw rgw ajt,

I marveled both at the way being just a little bit off at the starting point (in this case a fraction of an inch from the correct position) produced such a monumental error and at the remarkable coincidence of *this* particular typing blunder occurring with *this* particular passage. Happily, only a small adjustment was necessary to put things right again. The longer I would have gone on in this way, though, the more difficult it would be to correct.

This lesson occurred to me again and again while writing this book. Inaccuracies at the outset, if pursued long enough, can have enormous consequences. Even a small "system heading error," as it's called in navigation science, can, over some distance, lead travelers far from their original target. For every one degree off course a plane travels, for instance, it will miss its intended landing spot by 92 feet (28 meters) per mile flown. On a flight intending to go from New York to San Francisco, that one degree error would put passengers more than 50 miles (80 kilometers) from their intended destination. Bigger errors have bigger consequences for where we end up.

Common causes of ordinary navigation errors include imperfect knowledge of starting conditions, calculation mistakes, and tech failures. In a more metaphorical sense, these are also common reasons why people sometimes find themselves and their systems heading in the wrong direction or ending up in places they'd rather not be. At its core, this book is concerned with certain initial fundamental errors in our understanding of who and what we are as humans living in this world with others, and how progression along a course determined by those (along with the calculations based on them and technologies applied in service to them) is leading us far from where we'd like to be. Correcting course will entail, among other things, a recalibration of our basic knowledge of "starting conditions," the fundamental conditions we in fact are *always* working with. This book is intended to facilitate that process.

Featured here are some of the motes and misdirections which, especially in modern Western thought, have obscured our view. The main focus is on how these distortions have contributed to and constrained our abilities to combat the urgent global socio-environmental crises humanity confronts. Only in dealing with those subtle yet significant impediments can we begin to realize our full potential to address complex problems and participate with greater awareness in shaping our collective future. Time is not on our side, but the ace up our species' sleeve is the unique and invaluable capacity to learn to see the world differently and, with clearer vision, to move more deliberately in a different direction.

Relative to the current and future material suffering these crises bring, the ideas, concepts, and mental pictures that determine how we see the world may seem too amorphous to matter, too insignificant to waste precious time on. To the contrary, I argue, the paradigms they comprise direct our action, or inaction as the case may be, and shape our sense of what's possible in important ways. The good news is that, as with the typing error described above, the fix is surprisingly (even embarrassingly) simple. And we already have everything we need; it's just a matter of putting things together in the right way.

Why This Book

This project began with a relatively simple goal. It was meant to be a response to half a century, at least, of pleas for a better integration of diverse knowledge from across the disciplines, especially the natural and social sciences. In order to be able to use that knowledge to address the complex planet-scale problems human activities have brought about, scholars argue, we need a coherent and comprehensive way to organize it, preferably one that provides a more accurate mental picture of the interactions between people and their biophysical environments. When I first began to encounter these observations, I assumed that I was late to the party and that, surely, by that point others had stepped in to address those needs. I started searching for answers. Here and there were scattered attempts, yet the calls kept coming.

In books, journals, news, and online venues, I encountered the same basic message almost daily: *we need an integrative framework to make sense of the diverse, massive, and growing body of available information relevant to critical interconnected environmental and social issues*. The only thing that changed was that, at some point along the way, a second demand appeared. In response to more recent findings that massive and immediate transformations in social systems are necessary to avert the worst of social and environmental disasters, socio-environmental researchers looked to the social sciences for guidance on how to begin to make the needed changes. Dismayed by what they found, or more to the point didn't find, pleas for a reliable theory of social change became increasingly common, adding another layer to the conversation.

Meanwhile, it began to occur to me that the general sociological theory that had recently taken shape through a synthesis of important insights from sociology and other disciplines was actually, at the same time, a comprehensive framework for socio-environmental studies and a general theory of social change (Kasper 2016). In the pages that follow, I introduce that theoretical framework and show how it does triple duty, with two main goals in mind. In addition to answering the calls described above, this book also seeks to contribute to fundamental shifts in how we think. Of particular concern are prevailing views, in the modern Western context at least, of ourselves and our place in this world we share with countless others.

Without presuming that improved understanding fostered by a more adequate paradigm will automatically translate into rational problem-solving action, I maintain that it is an extremely helpful, if not necessary, condition. Until our understanding of these essential features of life more closely aligns with reality, we will remain trapped in our own confusion—unable to effectively respond to the crises we create and bound to keep generating the same kinds of problems again and again. Looking beyond the symptoms of these problems to confront their deeper roots, this book guides readers through a process of shifting their perspective and overall paradigm in the ways needed to better comprehend and work with the socio-environmental processes we're always engaged in, whether we know it or not.

What to Expect

To get the most of a book like this, it's helpful to have a sense of what to expect. With that in mind, I briefly describe some of the main qualities and features readers can anticipate.

Wide Ranging Subject Matter

The dynamic and complex confluence of different sorts of phenomena represented by the term "socio-environmental" makes this work inherently interdisciplinary. In my effort to create a sensible socio-environmental whole, I freely cross disciplinary borders, gathering relevant knowledge wherever I find it. Given the endless quantities of knowledge available, the biggest challenge was determining what to include. Regrettably, given the impossibility of knowing about or incorporating everything, a great deal will inevitably be left out. Final decisions about what to include were influenced primarily by consideration of what readers, with their radically differing background knowledge, would need to make sense of it all.

Diverse Target Audience

The reading audience I had in mind while writing was almost as eclectic as the subject matter itself. With the initial motivation coming from the observed needs of socio-environmental studies,[1] the book is aimed at scholars and teachers in that field (whatever discipline they might originally hail from) and the students they work with. At the same time, because the development of an integrative framework for socio-environmental studies depends so heavily on sociological knowledge and theory, and understanding the checkered history of both, the content also naturally speaks to sociologists, social scientists in general, and a subset of that population who are proponents of the work of German sociologist Norbert Elias (1897–1990), on whom the series this book is part of centers. More on him in a moment.

Across this spectrum, readers will almost certainly vary in familiarity with subject matter in the natural and social sciences, environmental issues, and background knowledge of all three. The consequent need to level the playing field informed the structure of the book, which aims to strike a balance between providing enough context to understand the problems and questions at stake without overwhelming readers with too much information. At various points, no doubt, some will wish for fewer details, while others may want elaboration. I humbly ask for readers' patience and understanding in these moments and hope that, for those wanting more, the bread crumbs provided are enough to set them on a path of further exploration.

The Remarkable Contributions of Norbert Elias

The writings and ideas of Norbert Elias, though not well known in American sociology and almost wholly unknown in socio-environmental studies, have been central to the development of my thinking about both. This is why they figure prominently in these pages, and why the book itself fits well within a series devoted to highlighting the influence and theoretical approach of Elias. For those unfamiliar with his writings, this book presents an opportunity to become acquainted with his wide-ranging, astute, highly relevant, and clearly stated insights. In doing so, they, like most who know Elias' work well, will likely come to appreciate its "considerable potential utility for combating the two-levelled, simultaneously practical and academic crises we humans have thus far faced, and are still currently facing" (Dunning and Hughes 2013:2). Highlighting the relationship between our academic and practical crises related to socio-environmental issues, this book is both inspired by and built on insights gleaned from Elias' writings.

I was first introduced to Elias around 1999 (nearly a decade after his death) as a young sociology graduate student in a contemporary social theory course. The circumstances of my intellectual development had primed me to profoundly appreciate what he had to offer. I was maturing out of that phase where I saw every author of assigned readings as an authority and took my job to be simply learning what they said. I had begun to take a more integrative and critical approach and, as a result, a certain dissatisfaction with the state of sociological thought was starting to creep in. At the same time, I was finishing my master's thesis, which focused on subject-objectdualism as a problem in sociology. My work in it was heavily influenced by the ideas of Merleau-Ponty (and other phenomenological thinkers), whose writings I had been studying intensively.

I have only vague memories of working my way through *Phenomenology of Perception* and other works dealing with related issues, but I vividly remember feeling a sense of elation and relief in encountering scholars who emphasized embodiment and the impossibility of disentangling the body from perception, thought, social life, and the world. It seemed like an antidote to the long-standing tradition in Western philosophy of placing immaterial consciousness as the source of all knowing. But I also remember those readings being fairly tough going. Having to communicate about these phenomena in the standard language of modern Western thought, I later realized, was no easy thing. So, when I first read Elias' treatment of these same issues, I was delightfully shocked by how clearly he was able to articulate them. How, I wondered, did he make it seem so straightforward and easy? The answer turned out to be simple. Elias took nature, the body, and material reality seriously and as the ground upon which any legitimate understanding of human social life must stand.

The introduction to *On Civilization, Power, and Knowledge* highlights moments from his early days as a student of both medicine and philosophy, when he recalled feeling dissatisfied with the discrepancy between the philosophical, idealist image of a person and the anatomical, physiological one (1998:3–4). This, we learn, was the beginning of his gravitating away from philosophy and toward a more processual, relational, and embodied view of persons and societies. The Eliasian smorgasbord served up in that book gave me a taste of the wide variety of fascinating subjects Elias dealt with and the potent insights that came out of his engagement with them. Those insights and the clarity with which they were expressed were what initially hooked me. It is that combination, along with the other rare qualities his work exhibits, which have kept me hooked ever since. I particularly appreciate Elias' insistence on telling it like it is (whatever the implications for an academic discipline may be), his consistent complementary use of the theoretical and empirical in social research, and the way he grounds sociological phenomena non-negotiably in the biological and physical contexts in and through which they occur.

Together, these and other aspects of Elias' overall approach point toward a highly practical synthesis of knowledge across disciplines. It is this synthesis which I endeavor to explore and develop here, drawing on the works of many and with Elias as a continuous thread throughout. So, while this book is not about Elias *per se*, it showcases the value of his work for helping us finally get beyond certain perennial and paralyzing problems in the social sciences and, it turns out, in socio-environmental studies, as well.

Existing Knowledge, Novel Presentation

Whether the information, ideas, and arguments presented here feel familiar or brand new, readers should know that, for the most part, they are not original to me. What is new is how they are put together. The possibility of the synthesis this work represents depends on the persistent labors of so many who shared their ideas, identified problems, and brought forth critical observations and arguments, even if unpopular, over the course of centuries. I am immeasurably grateful to them and for the opportunity to draw on aspects of their work to contribute, even if in a small way, to the advancement of our shared goals.

It's also worth noting a few key differences between this project and like-minded efforts. One is that it picks up where most others leave off.

Going beyond mere alerts to the crises of socio-environmental knowledge, this book ventures to offer some concrete next steps for getting past them. Also, in contrast to most sociological theory, the framework presented here is accompanied by a visual diagram (which serves as a kind of package to contain and convey what could otherwise be an overwhelming quantity of information) and supports for using it, thus facilitating communication, teaching and learning, and research in valuable ways. One final difference to mention here is its accessibility. This book is meant to be highly comprehensible, digestible, and useable for anyone with an interest in its themes. I avoid specialized jargon wherever possible—the kind that has thus far made it difficult to communicate across disciplines—introducing only two concepts likely to be new to most readers: *figuration and habitus*. They serve as alternatives to the more dualistic (at least as conventionally used and understood) concepts of society and individual. Devoting a full chapter to each (Chapters 6 and 7), I discuss the need for concepts which can better represent the reality of dynamic interdependence, the nuances of meaning which make them more appropriate, their place in the overall framework, and some of the practical applications of their use.

My efforts to produce an academic work that is inviting and easy to read have resulted in a kind of hybrid style which is simultaneously formal and down-to-earth. Apart from the introduction and conclusion (and rare other spots), my personal voice is minimized as much as possible. I do, however, freely use "we," "our," and "us" when talking about humans (which feels more natural than using "they," as if I'm not one) and, at other times in reference to Americans, Westerners, or academics. The context makes the reference group clear, avoiding the "royal we" syndrome. But even with these efforts to accommodate, readers should expect to exert themselves. Retraining habits of thought and redrawing mental pictures require a good deal of effort, time, and imagination. So, patience is advised in working through the chapters and any parts that may, on a given day, feel more challenging than others.

Summary of Chapters

Introducing the knowledge crisis, Chapter 1 tells the story of amassing ever-growing quantities of data, information, and knowledge studied within increasingly narrow specializations without an effective means of organization. These combined circumstances make it difficult to even know what we know, much less act on it to meaningfully address the urgent global crises confronting humanity. In spite of efforts toward interdisciplinarity to redress the problem, the situation has worsened and demand for a more integrated and actionable socio-environmental theory persists. In the face of this crisis, some have concluded that a new kind of specialist is needed, a synthesist, responsible for keeping the big picture in mind and maintaining a sense of what goes where. As it turns out, this demand goes back further than one might expect.

Chapter 2 explores the history of this demand, which became salient in the nineteenth century as specialization intensified, especially in the sciences. Achieving synthesis there proved impossible as long as the biological and physical sciences were believed to be irreconcilable. Eventually, certain shifts in scientific thinking allowed for the integration sought. It was only a matter of time, many believed, before the scientific study of human social life was brought into the fold, finally initiating the age of Progress. The expected forward momentum was stalled, however, by badly flawed paradigms for making sense of the social realm.

Chapter 3 examines the state of the social sciences, especially important to understand given the constant demand for their more robust presence in the study of environmental problems. After failed attempts to produce a human science that operated as successfully as physics, chemistry, and biology elicited skepticism about the value of the social sciences, social science disciplines doubled down on efforts to prove their relevance, but failed to critically reflect on the fundamentals. With a mindset of curiosity, this chapter explores basic questions about social science and especially sociology. In reviewing more than a century of disciplinary critiques, we discover some of the key conceptual hindrances to advancements in sociology and how a subtle shift in perspective—namely, the simple move of placing humans in their biophysical contexts—can help. I discuss the implications of this for the structure of social science disciplines and for the prospects of finally establishing an empirically-grounded general theory of human social processes.

In Chapter 4, we see how this framework functions as a map to guide us through socio-environmental territory, as yet relatively foreign to conventional thinking. The satellite perspective it takes allows us to observe, at a very high level of synthesis, relationships among different categories of phenomena operating at multiple levels. In doing so, we are able to derive sorely needed foundational premises for the sociological sciences and discern a general pattern at work in long-term socio-environmental processes. The following chapters focus, in turn, on each of the five components of this map, in the process elucidating and attempting to correct some of the most pervasive errors in our thinking.

One of the main challenges for socio-environmental studies has been the view, explicit or implied, that humans and nature are somehow separate. In exploring the concept of biophysical conditions, Chapter 5 supports our capacity to recognize distinctions between physical, biological, and social levels of phenomena without ascribing separation. The concept of biophysical conditions, as opposed to more conventional terms, attunes our awareness to the vast range of (largely invisible) conditions which have shaped, are shaping, and will shape us, our, societies, and the trajectory of life on this planet. In doing so, we cultivate the skills of noticing, navigating, and investigating the territory we are always in, but rarely paying attention to.

Acknowledging the biophysical conditions that humans are and operate within equips us to recognize the true nature of human societies, which are not *things*, but dynamic patterns of interdependence. Chapter 6 focuses on how grasping these inherent qualities of human associations resolves the supposed mysteries of the individual-society relationship and brings the social and environmental implications of increasing global interdependence into sharper view. Employing Norbert Elias' term "figuration" as an alternative concept, this chapter prepares us to retrain our minds to recognize, and be able to more accurately think and speak about, dynamic interdependence as a basic condition of human life.

Hand-in-hand with the modern Western view of societies as things is the notion of individuals as discrete, independent, self-existent entities. This illusion, which developed over centuries and is especially powerful today, lies at the heart of so many of the difficulties we have understanding and knowing how to approach socio-environmental issues. Chapter 7 discusses human persons as they really are: inescapably interdependent relational processes. I use habitus as an alternative term and a way to sidestep current conceptions of the individual in opposition to society. As used here, habitus represents the group(s)-specific "second nature" people necessarily develop growing up and living within particular figurational and biophysical conditions.

Although we've long understood that humans impact "the environment," our faulty mental models and concepts obscure our understanding of those impacts and who's at the receiving end, and have steered us largely toward ineffective "solutions." Chapter 8 looks at the mismatch between the magnitude of the problems we have created and the most common approaches to addressing them. Situated in the context of the socio-environmental synthesis framework, the concept of socio-environmental impacts provides further support for upgrading our habits of thought and serves as an important tool in the work of changing our impacts and identifying possibilities for intentionally bringing about the many other kinds of changes our situation calls for.

Change happens; there's no getting around it. At this time, though, it's crucial that we better understand change processes to gain insights into two critical questions: how to make the changes deemed necessary to avert catastrophic collapse and how to best navigate the disruptions to business-as-usual that are coming, no matter what we do. At present, we are not well equipped to answer either question. This weakness has garnered increasing attention of late as our windows of opportunity to forestall climate and other sorts of disaster close. Chapter 9 discusses how the socio-environmental synthesis framework serves as an actionable theory of social change.

The conclusion brings us back to the theme of synthesis in socio-environmental studies, and how enhancing our ability to see and make use of available knowledge can aid our efforts to navigate increasing turbulence and guide social change in ways that minimize harm and maximize health and happiness. This section explores potential applications of the socio-environmental synthesis framework, especially in education and research, and highlights some of the most fundamental changes said to be necessary for humans to survive and thrive beyond these crises. Ending where we began, the conclusion of the book emphasizes the need for a clear-eyed and more reality-congruent sense of ourselves and the world. Only with this can humanity hope to avert catastrophic systems collapse, craft lives together with greater awareness, and more intentionally shape our collective future, rather than obliviously letting it just happen *to* us.

Note

1. This area is most often referred to as environmental sciences or environmental studies, but because the issues of greatest concern are caused by people and affect people, they are inherently social. Making that explicit in our most basic labels is an important step toward establishing a mind-set and academic structures capable of containing both dimensions.

References

Dunning, Eric, and Jason Hughes. 2013. *Norbert Elias and Modern Sociology.* London: Bloomsbury.

Elias, Norbert. 1998. *On Civilization, Power, and Knowledge,* edited and with an introduction by Stephen Mennell and Johan Goudsblom. Chicago and London: University of Chicago Press.

Kasper, Debbie. 2016. "Re-conceptualizing (Environmental) Sociology." *Environmental Sociology* 2(4):322–332.

Keble, John. 1851. *Lyra Innocentium: Thoughts in Verse*, 5th ed. London and Oxford: John Henry Parker.

Contents

List of Figures

CHAPTER 1

The Crisis of the Librarian

The greatest crisis we face, science fiction writer Robert Heinlein argued in the middle of the twentieth century,

> is not Russia, not the Atom bomb, not corruption in government, not encroaching hunger, nor the morals of the young. It is a crisis in the organization and accessibility of human knowledge. We own an enormous 'encyclopedia' which isn't even arranged alphabetically. Our 'file cards' are spilled on the floor, nor were they ever in order. The answers we want may be buried somewhere in the heap, but it might take a lifetime to locate two already known facts, place them side by side and derive a third fact, the one we urgently need. Call it the Crisis of the Librarian. (1952:21–22)

Given the many urgent planet-scale problems we face—global climate change, mass extinction, and persistent social conflict, to name just a few—it may seem strange to call the organization and accessibility of human knowledge our "greatest crisis." If the world as we know it is ending, is it really appropriate to get all worked up about flaws in our information-organizing systems? Heinlein thought so.

He didn't get everything right in his prognostications about the then-mythical year 2000, but he was uncannily alert to the crisis of knowledge. Before computer, internet, and portable wireless technologies came to define our way of life, there was already a widespread sense that something

D. Kasper, *Beyond the Knowledge Crisis*,
Palgrave Studies on Norbert Elias,
https://doi.org/10.1007/978-3-030-48370-8_1

was wrong. Despite the massive amounts of information and knowledge humans have accumulated at ever-increasing rates, and despite our growing abilities to manipulate physical and biological processes for our own ends, we seem to be relatively incapable of dealing with complex social problems, as evidenced by serial wars, ongoing violence, widespread deprivation, and now, global environmental catastrophe.

Assuredly, the trouble is not that we don't know enough. We know more than ever about how Earth systems work and the havoc human activities are wreaking in them. We have a pretty good sense of what's necessary to mitigate some of the worst outcomes for humanity and other life forms. We have loads of data about really specific things too, like how much carbon dioxide particular nations, activities, and appliances emit each year on average, rates of increase in ocean acidity, which species are likely to be directly and indirectly affected by certain environmental changes, and so much more. In short, even with all the usual caveats about scientific knowledge being provisional, we know an awful lot.

The current fragmentation of knowledge, however, makes it difficult to even *know* what we know, much less act on it in meaningful ways. In particular, we lack a clear sense of how to implement the changes we know are needed. This simultaneous knowing and impotence is the real tragedy, the ultimate crisis. If there is to be any hope of overcoming it, we need a more workable system for organizing, accessing, and using what we know. In a nutshell, that's what this book is about. But before we can move beyond the crisis of abundant yet disconnected specialized knowledge, it's helpful to have a sense of how we got here in the first place.

Toward Specialization: A Brief History

From the earliest days of formal education, some way of dividing scholarly labor appears to have been indispensable. The curriculum in ancient Greece, for instance, was organized into the language-oriented *trivium* (consisting of grammar, logic, and rhetoric) and the math-focused *quadrivium* (comprised of arithmetic, geometry, music, and astronomy). By the late medieval era, these traditional Greek courses of study had become preliminary for education in the professions of medicine, law, and theology. Still, up to this point, the emphasis was on establishing and using a general method of academic inquiry, rather than on discovering new knowledge (Dirks 1996). This was to change in the coming centuries.

The fifteenth, sixteenth, and seventeenth centuries brought major shifts in how people viewed the universe and sought knowledge about natural phenomena. Intense scientific activity and associated changes in social thought ushered in what is often called the Enlightenment era, a period of ongoing discovery spanning the mid-1700s to late 1800s. Especially significant was the establishment of a scientific method which prioritized empirical evidence, the role of mathematics, and the accumulation of new knowledge. As humanity's store of scientific knowledge about the world grew, so did the need to organize it. Thus commenced a period of more intensive specialization in knowledge about particular aspects of reality.

The fact that specialization occurred is not surprising. Less inevitable, though, were the specific forms it took in the academy. One obvious manifestation of this increasing narrowness was the structure of academic disciplines, which gradually took shape within particular social and historical circumstances. Noteworthy among them were the professionalization of scholarly activity (which was formerly viewed as more of a natural gift or religious-like vocation) and the creation of institutional structures in the form of academic departments, journals, societies, and reward systems (Dirks 1996). This is particularly evident in the history of the American university.

Today it is nearly impossible for college students in the United States to imagine their schools without the departments and majors that seem so essential; they might be surprised to learn that this setup was far from normal in the nation's young institutions. In 1765, for instance, the College of Philadelphia (now the University of Pennsylvania) became the first among the nation's nine chartered universities and colleges, and most of the non-chartered ones, to institute the co-existence of more than one department when it added a medical school to the existing "collegiate." It was not until 1825 that Harvard created departments, with great resistance from its faculty, and the University of Virginia opened with parallel curricula in seven different departments they called "colleges."

Following the US Civil War and the combination of discouraging enrollment trends, a desire to remain competitive with European universities, and an increase in available surplus wealth, a period of deep reform in higher education began. Tensions ran high as a result of competing ideas about the purpose of American higher education. While some advocated a unity of knowledge and cultural standards through the liberal arts, others prioritized research and the development of general empirical methods for solving particular problems, and yet another faction viewed practical public service through vocational training as the proper goal of higher education, one that would bridge the gap between lofty academic pursuits and "real life" (Vesey 1965). On top of these competing interests, educational reformers of all stripes had to contend with hostility to higher education from certain sectors. Industrial leaders and ill-educated Americans, in particular, had "little enthusiasm for the foreign, the abstract, or the esoteric" and expressed a general mistrust of bookishness and a skepticism about the material security that an education could bring (Vesey 1965:13).

The eventual success of universities and colleges in the United States came less from winning over an unsympathetic populace than from favorable political circumstances and strategic maneuvers through which promoters of education were able to incentivize the creation of universities across the states. Around 1890, as this period of intense reform was drawing to a close, it seemed that proponents of a unified academic culture, specialized research, and public service were peacefully coexisting, both within and between institutions. This tenuous harmony, however, would soon evaporate as education's pendulum began to swing in a different direction.

The intellectual wholeness fought for by liberal arts advocates gave way to narrower elective courses of study. Interestingly, champions of research were also critical of this shift, arguing that it led to an excessive smattering, faulty preparation, and cultural amateurism. At the same time, science itself was becoming an object of contention. The word science went from referring to an organized body of information about a subject to an approach to knowledge-seeking that aspired to account for the entire universe. This shift inspired mistrust, especially by those who drew clear boundaries between Nature and Spirit. Ultimately, these tensions resulted in a series of fractures.

Psychology, with a new emphasis on scientific research, split off from the more idea-based philosophy. English was divided into different concentrations—one on culture, via literature, and the other on more empirical philological research. Sociology diverged from economics, which soon underwent its own internal division. Meanwhile, the very idea of topical specialization was reified into rigid academic divisions with the establishment of professionalized disciplinary training in graduate programs.

The first two decades of the twentieth century brought an intense splintering of new fields from existing disciplines. New departments proliferated, departmental expansion became an end in itself, and competition for resources to support it became the norm. Prominent academics began to voice concerns about over-specialization and the loss of cohesion in higher education institutions (Moran, J. 2010). Universities soon "crystallized into a collection of divergent minds, usually ignoring each other, commonly talking past one another, and periodically enjoying the illusion of dialogue on 'safe' issues" (Vesey 1965:58).

Interestingly, this description of turn-of-the-twentieth-century higher education could easily be mistaken as contemporary. With increasing specialization being the dominant trend in higher education for the past hundred years, academics today still share the same concerns (Gaff and Ratcliff 1997). The only real difference is that, after moving for so many years in this general direction, we find ourselves much farther down the road.

Fragmentation and the Knowledge Crisis Today

In stark contrast to the early days of American higher education, when introducing a new special degree was considered radical, by the first decade of the twenty-first century the Digest of Education Statistics listed over 1000 different degrees that could be obtained in the United States at the bachelor's, master's, or PhD level—*not* including professional degrees like law or medicine or the many specialized associates degrees available. Within the 30 major headings the list is divided into, one finds further specialization in respective fields: 86 separate specialized degrees under business, for example, 96 in education, and 182 in health professions and related clinical sciences (Jacobs 2013:205). In terms of degrees offered, applied fields exhibit greater levels of specialization than the arts and sciences, but even within identical degrees in those fields, growing numbers of subdisciplines reflect high levels of specialization.

In my field of sociology, for example, the American Sociological Association (ASA) currently recognizes 52 subsections, representing the many specialties with which fellow PhDs might identify. This is up from two in 1959, five in 1961, and 44 in 2006. According to the ASA Section Manual, growth has been consistent, but especially intense in the past two decades. Other disciplines show similar trends. At the time of this writing, the American Political Science Association (APSA) is split into 47 sections and the American Psychological Association (APA) has 54 divisions. In many ways, this differentiating makes sense. There are, after all, real constraints on the breadth of mastery one can achieve. Among them is the challenge of keeping up with the research in a given field, a challenge that has grown dramatically in recent years as information technologies have expanded availability and access.

Today, we are literally surrounded by information. It takes not only the familiar shape of letters and numbers imprinted on readable surfaces in any number of ingenious formats, but we are awash in a steady stream of waves bouncing here and there, transmitting the ones and zeros which are ultimately decoded by computers and appear to us as words, images, and sounds. As the encyclopedias and file cards Heinlein mentions have come to be replaced by search engines and databases, the magnitude of the knowledge crisis has only grown. In our so-called information age, the simple effort to learn something about a particular subject can quickly induce a sensation of drowning. Trying to find information today has been likened to attempting to drink from a fire hose. So as to not be overwhelmed by its volume, speed, and force, the torrent must be diverted into smaller courses, reducing it to a flow that the human organism can usefully assimilate. It is not surprising, then, that growing specialization accompanied the growth of information and knowledge.

By 1970, specialization was said to be "doubling every decade or two," as measured by the multiplication of professional organizations, journals, courses within respective fields, and the classification terms used in abstracting and indexing services (Lasswell 1971:440). In fact, the growth rate of active peer-reviewed scholarly and scientific journals has been an almost constant 3.46% per year for most of the last three centuries (Mabe 2003:193), representing a doubling about every 20 years. In the decade between 2002 and 2011, the number of active, refereed, scholarly journals more than tripled, going from 16,925 in 2002 to 57,736 in 2011, while at the same time, the number of articles per journal and pages per article increased at even higher rates (Tenopir and King 2014:167).

All of this points to a massive explosion of scholarship that, in principle, academics might be expected to stay informed about. In practice, though, keeping up even within just one discipline became virtually impossible. In this context, specialization among and within disciplines can be seen as a useful tool for cordoning off one's attention to allow focus on a particular slice of reality. This way of thinking about it sheds light on the meaning of the term "discipline"—as in a discipline of focus on *this,* but not *that*.

The problem is that, without a complementary effort to pull it all together and organize it in some way, disciplinary isolation only intensified. Boundaries were solidified and reinforced through specialized terminology and advanced training, exacerbating the fragmentation of knowledge. Rather than a functional adaptation, the resulting disciplinary structure came to be increasingly viewed as a hindrance to communication across fields, the growth of genuine knowledge, and our ability to deal with complex problems. This "spread of specialized deafness," as Kenneth Boulding put it, "means that someone who ought to know something that someone else knows isn't able to find it out for lack of generalized ears" (1956:199). As a consequence, there was growing recognition of the problem of "undiscovered public knowledge"—knowledge, which in its fragmented state, remains unrealized because the necessary pieces of information have not been put together (Swanson 2001). This is precisely the situation Heinlein saw as "the greatest crisis facing us."

One of its greatest dangers lies in the ways that the fragmentation of knowledge can misrepresent and distort reality. How we currently talk about the "natural sciences" and "social sciences," for example, gives the impression that their subject matter—biophysical and human social phenomena, respectively—somehow exists separately. This has served to reinforce the modern western view of humanity as a phenomenon alongside of or above nature, as opposed to *in* and *of* it. Further obscuring reality are the politics of academia, where jockeying for prestige and resources impels each discipline to promote its distinctive significance at the expense of understanding the relationship between its subject matter and that of other disciplines.

The resulting sense that economics and ecology, for instance, are merely *different* areas of inquiry obscures their interrelatedness in ways that do real harm. This separation lends itself to a view of "the economy" as an independently existing thing, encouraging us to think about the production and profits related to cars, computers, corn, and cancer treatments in abstract terms (e.g., Gross National Product) which ignore their biophysical contexts and socio-environmental consequences. From these critical observations of the ways that disciplinary isolation can misrepresent reality have come various efforts to set things right. Prominent among them was the push for interdisciplinarity.

Re-uniting Knowledge

In a 1993 MIT lecture, physicist John Armstrong colorfully reminded his audience that "God did not create the universe according to the departmental structure of our research universities." Whatever the practical utility of disciplinary boundaries might be, he subsequently wrote, "they lull us into forgetting that nature is interconnected and complex in ways we still only dimly perceive" (1994:118). The rise of interdisciplinarity reflects the desire to pursue a more unified form of knowledge which more closely resembles a unified Nature. More practically, though, it was also seen as an antidote to the fragmentation of knowledge wrought by over-specialization and the solution to "problems and issues that cannot be addressed or solved within the existing disciplines" (Moran, J. 2010:13–14). If the structure of the academy must be changed to accommodate reality, like-minded thinkers argued, then perhaps the structure itself is part of the problem (Caldwell 1983).

These concerns catalyzed the formation of a number of interdisciplinary fields—beginning with area studies, American studies, and comparative literature in the 1930s and 1940s and expanding into the multitude of new "interdisciplines" like molecular biology, cognitive science, and biomedical humanities in the early twenty-first century. The rapid expansion of interdisciplinarity represents "a major episode in the history of knowledge," and the fact that the formation of new disciplinary combinations is no longer unusual is an indication that interdisciplinary studies serve a need that is not going away (Klein 2005:77).[1] Among the other evidence of growth in interdisciplinary activity includes: the appearance of interdisciplinary initiatives on campuses (Borrego et al. 2014), the prevalence of interdisciplinarity in new journals (Jacobs 2013), the multiplication of funding opportunities aimed explicitly at promoting interdisciplinary research (Lyall et al. 2013), and the increasing frequency of work attempting to bridge the natural and social sciences (Braun and Schubert 2007). This last item has been especially prominent in efforts to understand human-environment relations in order to be better able to address problems of increasing concern.

Toward that end, the National Academy of Sciences declared interdisciplinarity to be an integral feature of scientific research. They cite the inherent complexity of nature and society, the desire to explore problems and questions not confined to a single discipline, the need to solve complex societal problems, and the motivation to develop new technologies as reasons for the shift (2005:30–39). Movement in this direction is also evident outside of the sciences proper. The emergence of sustainability (with its explicit focus on the need to harmonize environmental, social, and economic systems) as a priority for cities, businesses, and organizations; a growing emphasis on resilience, and the normalization of the concept of "socio-environmental" as a more accurate descriptor of the interrelated social and biophysical subsystems that mutually influence one another are all examples. These developments are important for getting us out of the conceptual trap of thinking and speaking about environmental problems as if they were somehow separate from people and for reminding us that the problems observed are rooted in the relationships between human activities and biophysical systems.

Seeking Socio-Environmental Approaches to Socio-Environmental Problems

In the face of overwhelming evidence that human activities are endangering certain natural systems and thus jeopardizing human health, quality of life, and the long-term survival of our species, efforts to better understand the dynamic interdependence between biophysical and human social systems, practices, and behaviors have multiplied. In the 1960s and 1970s, novel environmental sub-disciplines, such as environmental psychology, environmental history, environmental philosophy, environmental sociology, and more were created to address the observed neglect of the environment in those fields. At the same time, arguments about the inadequacy of a discipline-centered approach were becoming ubiquitous in the literature.

These observations inspired attempts to re-imagine education and research along socio-environmental lines. Environmental sciences and studies, for instance, grew out of the "misfit between perceived need, experience, information, and the prevailing configuration of knowledge embodied in the disciplinary organization of academia" (Caldwell 1983:249). As the sense of this misfit expanded, so did environmental sciences and studies programs.

In the 1950s, Syracuse University was the first to institute an environmental studies program, awarding the first bachelor's degree in Environmental Studies in 1956. In 1965, Middlebury College in Vermont created the second program in the United States. From two in 1965, the number of interdisciplinary environmental studies programs quickly multiplied, with 90% of them at the undergraduate level. The total grew to more than 500 in 1990, and then more than doubled over the next twenty years to 1200. This record, and the continuing creation of new programs, made environmental studies and sciences "one of the fastest-growing fields of undergraduate study in the country" (Maniates 2013:256).

Political scientist Harold Lasswell described the rapid development of interdisciplinary environmental studies as part of a "counter offensive" to the growing fragmentation of intellectual life and decreasing numbers of scholars attending to "the map of knowledge as a whole" (1971:439). Others noted that this fragmentation was evident even in places where the purported goal *was* to promote holistic knowledge. Wes Jackson, botanical geneticist and Chair of one of the United States's first environmental studies programs, contends that the very existence of environmental studies highlights the failure of the liberal arts to muster the attention and cooperation needed to study big, complex, socio-environmental problems within standard departmental structures (Jackson 2013).

If the existence of environmental studies is an indicator of failings in higher education, then the inability to effectively integrate disciplinary knowledge reflects the failings, to date, of environmental studies. A 2010 assessment of college and university environmental studies and sciences programs concluded that they generally suffer from unclear goals, a dis-integrated disciplinary hodgepodge, and incoherent curricular smorgasbord (Clark et al. 2011). At a time when the need for dynamic interdisciplinary environmental programs has never been greater, "those who plan and deliver these programs appear to be selling their students and the planet short" (Maniates 2013:255).

Despite the best efforts of those of us teaching in environmental studies, effective integration has not been adequately achieved, and demands for a synthesis of relevant knowledge persist. Acknowledging that there is still a place for the separate study of physical, biological, and social systems, environmental psychologist Paul Stern (2013) argues that what is new and noteworthy is that "the space where it no longer makes sense to study the systems in mutual isolation has been growing at an accelerating rate." It is this very change, he says, that has long incited demands for a new science which studies the relationships between social and biophysical systems. One such demand came in the form of a 1993 paper calling for a "second environmental science" of human-environment interactions (Stern 1993). To the chagrin of sociologist, Riley Dunlap, this call came 15 years *after* the founding of environmental sociology, intended to do just that. For him, Stern's call and the fact that alternative approaches "like 'coupled human-natural systems' and 'sustainability science' and the like emerged without any seeming awareness of environmental sociology" reflect the silo nature of academia (Dunlap 2013).

Despite these developments and decades of intense efforts to integrate knowledge about the biophysical world with what we know about the human systems, activities, and behaviors which are part of and interact with it, researchers continue to churn out the same conclusion: *we need to do better*. In particular, the mantra goes, we continue to need a more effective integration of the social and natural sciences (Moran, E. 2010; Smith 2009; Stafford et al. 2010; Stock and Burton 2011; Tahir 2009; Victor 2015; Wei et al. 2015; Zax 2009). In taking a closer look at what we know about the planet, our current situation, and our prospects for the future, we can better appreciate why this integration is so very crucial.

Thresholds and Trajectories

As a result of generations of hard work observing, measuring, and gathering information from around the world, scientists have attained an impressive understanding of how non-human natural systems work and about the state of things within them, now and at various points in the past. More recently, they have also come to understand some of the key changes humans have brought about in these systems, from local to planetary scales. Especially significant are the critical tipping points we are now able to discern. Certain human activities have brought us to thresholds beyond which systems begin to behave differently, where the same activities set in motion mutually reinforcing feedback loops propelling a system onto a new course. The collapse of ocean fisheries, acceleration of melting ice sheets, upwelling of warmer ocean waters, methane release from thawing sea beds, climate volatility, extreme drought, fertilizer-induced shifts in lake ecologies, and the demise of tropical coral reef systems are all examples of troubling thresholds that have been well-studied (Rockström 2015). In all of those cases, anthropogenic impacts threaten the homeostasis of the systems on which humans, and countless other species, rely.

In order to create a more detailed picture of human-driven changes to Earth systems, a group of scientists assembled decades of data collected by the International Geosphere-Biosphere Programme—an initiative launched in 1987 "to coordinate international research on global-scale and regional-scale interactions between Earth's biological, chemical and physical processes and their interactions with human systems" (IGBP 2016). This group examined trajectories of a number of key indicators between 1750 (the start of the industrial revolution) and 2000. Their

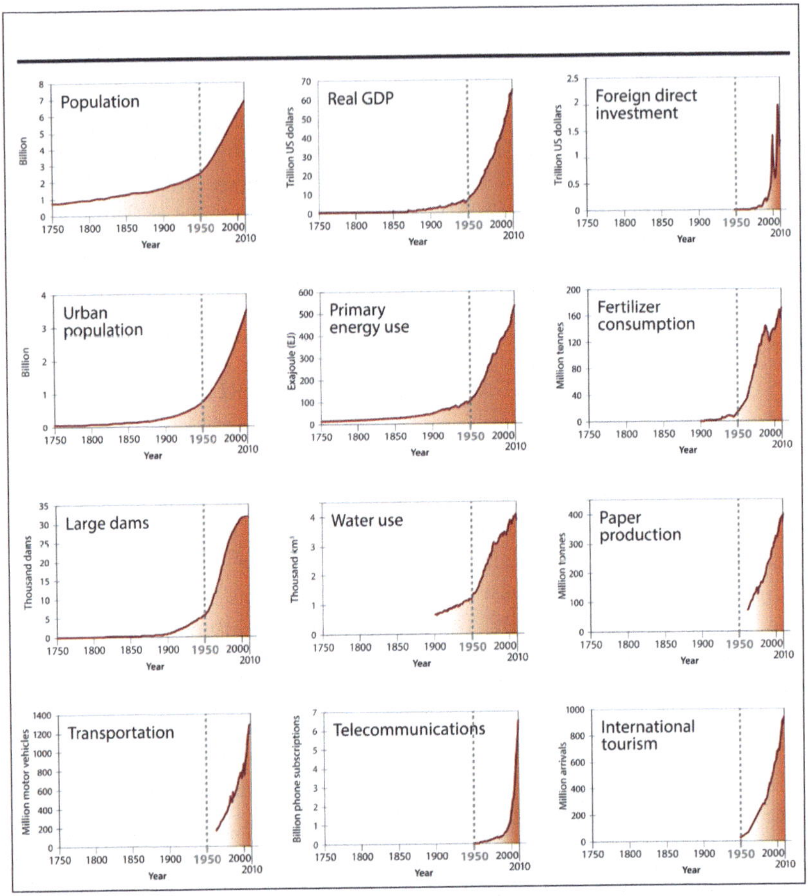

Fig. 1.1 Socioeconomic trends, 1750–2010 (Steffen et al. 2015:4) (Image used with permission of Sage Publishing)

efforts produced what have become known as the "Great Acceleration" graphs, named for the dramatic acceleration in Earth system and socioeconomic trends around the middle of the twentieth century, subsequently updated with data through 2010 (see Figs. 1.1 and 1.2).

Most striking about these graphs is how well the post-industrial rise of key production and consumption activities tracks with indicators of dangerous trends in Earth systems. Observing this offers a planet-scale

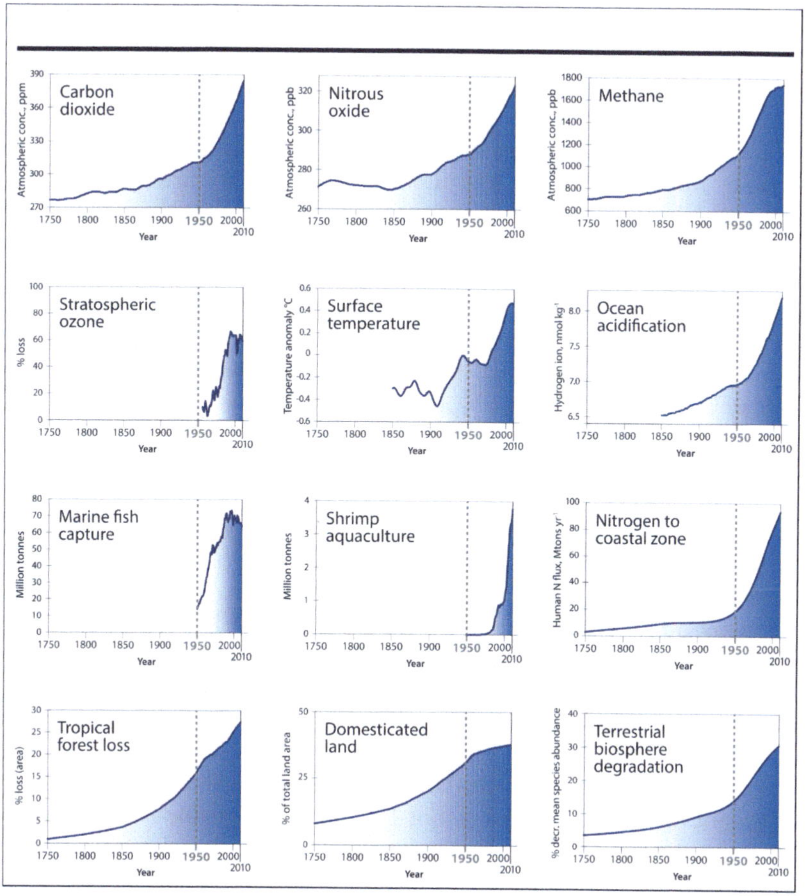

Fig. 1.2 Earth system trends, 1750–2010 (Steffen et al. 2015:7) (Image used with permission of Sage Publishing)

view of what many socio-environmental researchers refer to as the "coupling" of socioeconomic systems and the overall "Earth System," thanks largely to intense globalization of production and consumption practices. Related to this is an even more comprehensive threshold scientists have recently identified: the point at which humans became a force of change on par with the geo-chemical-physical forces which have thus far shaped the biosphere. This new epoch has been dubbed the Anthropocene.

The efforts which produced the "Great Acceleration" graphs not only provide evidence of this shift—showing that most Earth System indicators have moved "outside of the Holocene envelope of variability" observed over the past 12,000 years—but also pinpoint the onset of the Anthropocene as the mid-twentieth century. More precisely, they've designated Monday July 16, 1945, as the day we crossed that threshold (Steffen et al. 2015:12). In just over two generations—within the span of a single lifetime!—humanity has become a fast-acting force of planetary change. At the moment, the changes we're bringing about do not bode well for humans or the countless millions of species with whom we share the Earth and on whom we depend. For this reason, the most prominent theme in socio-environmental literature today is: *we can't go on doing what we're doing*. Most analysts put it more bluntly, stating that "business as usual" spells almost certain disaster. This is especially apparent in the phenomenon of climate change.

Climate science shows that the continuation of current trends in carbon dioxide emissions will soon carry us across dangerous system thresholds, beyond which planetary warming will accelerate regardless of what we do, making much of the planet uninhabitable for humans and many other species (Anderson and Bows 2011; Hansen et al. 2013; Stern 2006). Additional dangers noted are threats to national security (Steinbruner et al. 2013) human health (CDC; WHO), and societal collapse (Conniff 2012; Kolbert 2005). While the details of these prognoses and the treatments prescribed are contested (not surprising, given the complexity of the problems and the degree of economic investment in "business as usual"), what is not legitimately arguable is that human organisms depend on a very particular range of biophysical conditions for their survival. Simple logic, then, suggests that there are definite limits to what we can get away with doing to the biosphere.

Clinging to the "dominant western worldview" (Buttel 1996), characterized by its failure to acknowledge the biophysical bases of human social life, are some who willfully ignore or contest the very idea of limits. For example, economist Julian Simon famously took issue with the notion of "finite," wondering: "Why shouldn't the boundaries of the system from which we derive resources continue to expand…just as they have expanded in the past?" (1981:49). But as the Great Acceleration Project and other studies emphasize time and again, the recent human past is an anomaly in the larger picture of human history and thus a poor indicator of what future trends might look like. Looking back, we find our ecological limits consistently revealed in the collapse of societies who exhibited sustained disregard for them. Though aware of only a small fraction of the societies who have come and gone in this way, we now understand that societal collapse is more the rule than the exception (Tainter 1988).

As sociologist William Catton eloquently stated in his book *Overshoot*, "we are in no way protected from the consequences of our actions by remaining confused about the ecological meaning of our humanness, ignorant of ecological processes, and unmindful of the ecological aspects of history" (1982:vii). Though confusion and even ignorance are perhaps understandable, neither state alters the facts of our existence or the consequences of our actions. Allowing ourselves to remain in those states is pure folly, the kind that Blue Oyster Cult famously sang about in their 1977 hit song, Godzilla: "history shows again and again how nature points out the folly of men."

Given the multitude of limits we are currently courting and surpassing, the end of the world as we know it has become a popular topic. Though no one can predict exactly how a collapse scenario would unfold, the recent boom in speculative fiction about the unraveling (and sometimes re-organization) of society has plenty of visions on offer. Collapse scenarios in this genre and its various flavors (e.g., dystopian fiction, "cli-fi," and "prepper porn") usually involve some combination of themes related to climate change, post-fossil fuel, political-economic upheaval, and massive epidemic-induced die-off. They seek not so much to predict the future, but to elucidate important features of the present that, if continued, are likely to result in some form of brave new world.

This kind of fixation on "the end" is itself not new. Every modern generation has believed itself to be at the precipice of transition, argues author Nathaniel Rich. But, he says, things really *are* different now. "There has never been a generation in the history of human civilization with more access to bad news than ours" (Rich 2013). Endless information about global warming, infectious disease, food safety, nuclear warheads, widespread poverty, cybersurveillance, natural disaster, and more is only a few taps away. "We're information-saturated," Rich continues, "and most of the information is ominous." Among those paying attention, there is an anticipatory undercurrent of doom or radical social change or both. Dramatic flourishes aside, socio-environmental researchers confirm that these folks are onto something.

Turning Things Around

Analysts who understand the situation know that massive social changes are inevitable, for at least three main reasons: "the era of cheap and easy fossil fuels is over…climate stability is now a thing of the past…[and] we've reached the end of economic growth as we've known it in the U.S." (Miller and Hopkins 2013:i). More specifically, with the understanding that excessive warming (i.e., a couple of degrees above the pre-industrial average) is likely to result in widespread food, water, weather, and sea level stresses, climate scientists urge that "a planned economic contraction to bring about the almost immediate and radical reductions [is] necessary" (Anderson and Bows 2011:41). Admittedly, these actions are outside the bounds of conventional politics in most countries.[2] Yet the science repeatedly confirms that gradual or incremental changes, much less "business as usual," are a path to catastrophe. Rather, "to slow down, let alone reverse, increasing carbon emissions and temperatures requires the total reorganization of social life" (Urry 2009:198) and a "revolution in the sense of massive social movements and alterations in how we run our economy and use energy and resources" (Foster 2012). Summing it up concisely, writer Brentin Mock (2015) says, "want to fix the climate? first, we have to change everything."

The key takeaway here is that, ready or not, change is coming. On the one hand, doing nothing puts us on a path toward massive unplanned changes (especially in energy, climate, and economic systems), leaving us to react to crises as they unfold. On the other hand, making the necessary adjustments ahead of time in energy and economic systems to avert the worst of potential disasters entails sweeping changes of its own, albeit the kind made deliberately. Seeing this latter option as preferable, many researchers and concerned citizens have long advocated for proactive restructuring of key social systems to steer us away from widespread catastrophe, and toward greater resilience and more satisfying lives.

Generic prescriptions for such change abound. We are told, for instance, that "sustainability demands changes in human behavior" (Fischer et al. 2012:153), that "consumer cultures will have to be re-engineered" (Assadourian 2013:113), and that it will take "massive political will to counter the momentum of dangerous trends" (Raskin 2014:1). In short, there appears to be nearly unanimous agreement among socio-environmental researchers that a transition to sustainability "requires a social avalanche of unprecedented proportions" (Fischer et al. 2012:158).

The problem is that, while they have a lot to say about *what* should be done, they have much less to offer when it comes to the *how*. It is common for discussions like this to conclude by asking "who or what might start this avalanche?" (Fischer et al. 2012:157) and "who will change the world?" (Raskin 2014:6). With regard to a viable theory that can guide us in making the much-touted preemptive social transformation, sociologist Robert Brulle (2012) points out, "we really don't have one."

The Need for a Synthesis of Knowledge

In response to observations like these, others have noted that, in addition to ecological and social crises, we have a profound crisis in the dominant paradigms of environmental social science, which "can merely define a problem but offer no coherent and politically plausible program of action" (White 2012). Despite vibrant activity in the areas devoted to better understanding human-environment relations, we are still in the position of wondering how to best think about and study them, and increasingly asking how to actually *do* what it appears we must. Professor of sociology and human geography Karen O'Brien says, "we are still pretty much in the Dark Ages when it comes to taking in existing and emerging understandings of human development and social change." The hope that people will adopt universal values to catalyze change at the scale and rate needed is unfounded in a situation where science admonishes humans for crossing dangerous thresholds, but provides no sense of how we should respond. Our own disciplinary barriers, she adds, get in the way of bringing together the knowledge we need for "navigating the world back into a safe operating space" (O'Brien 2015).

It has become clear, as Jacobs (2013) argues in his defense of disciplines, that interdisciplinarity is not synonymous with integration or synthesis. Despite half a century of concerted interdisciplinary development in socio-environmental studies, the need—among practitioners, teachers, and students alike—for theoretical integration persists (Proctor et al. 2013). The last decade has illustrated that "integrating research disciplines to deal with complex sustainability related problems is far from unproblematic," and that there remains a wide and stubborn gap between the ideal of disciplinary integration and its achievement in reality (Stock and Burton 2011:1092).

Increasing fragmentation of, and competition among, a growing mix of disciplines and methods, along with the inability to effectively communicate across them, impedes the integration of the scientific work necessary for meaningfully addressing our increasingly urgent socio-environmental problems. And even if we could successfully integrate and apply all of the relevant knowledge we have, there's no guarantee that we could turn things around. Then again, *not* trying pretty much guarantees that we won't. What have we got to lose? The question then becomes: How can it be done?

Following his naming of the crisis of knowledge, Heinlein proposed that, in order to integrate knowledge in the ways necessary, "we need a new 'specialist' who is not a specialist but a synthesist" (1952:22). This is the idea explored in the next chapter. In examining the history of demands for a synthesis of knowledge, and some successful and failed attempts to meet them, we gain valuable insights into how to achieve the socio-environmental synthesis so desperately sought.

Notes

1. As new areas of study developed, so did a new vocabulary of terms to distinguish between varying degrees of disciplinary involvement, including, *cross*-disciplinary, *multi*-disciplinary, and *trans*-disciplinary, in addition to *inter*-disciplinary. For simplicity's sake, I use the most common term, interdisciplinary, here to refer to diverse efforts to work across disciplinary lines. For present purposes, I tend to side with Rick Rylance in finding "this faintly theological hair-splitting unhelpful" (2015:314).
2. Interestingly, the drastic measures taken to prevent the spread of novel coronavirus (COVID-19) in 2019 and 2020 offer an important precedent showing that what was previously unthinkable—e.g., halting certain economic activities, curtailing freedom of travel, canceling major professional and collegiate athletic events, internationally coordinated efforts—is possible and can be achieved in a rather orderly and peaceful manner.

References

Anderson, Kevin, and Alice Bows. 2011. "Beyond 'Dangerous' Climate Change: Emission Scenarios for a New World." *Philosophical Transactions. Series A, Mathematical, Physical, and Engineering Sciences* 369(1934):20–44.

Armstrong, John A. 1994. "Is Basic Research a Luxury Our Society Can No Longer Afford?" *Advanced Materials and Processes* 145(5):117–121.

Assadourian, Erik. 2013. "Re-engineering Cultures to Create a Sustainable Civilization." Pp. 113–125 in *Is Sustainability Still Possible?*, edited by Worldwatch Institute. Washington: Island Press.

Borrego, Maura, Daniel Boden, and Lynita K. Newswander. 2014. "Sustained Change: Institutionalizing Interdisciplinary Graduate Education." *Journal of Higher Education* 85(6):858–885.

Boulding, Kenneth E. 1956. "General Systems Theory—The Skeleton of Science." *Management Science* 2(3): 197–208.

Braun, Tibor, and Andras Schubert. 2007. "The Growth of Research on Inter- and Multidisciplinarity in Science and Social Science Papers, 1975–2006." *Scientometrics* 73(3):345–351.

Brulle, Robert J. 2012. American Sociological Association Environment and Technology section listserv, November 17.

Buttel, Frederick H. 1996. "Environmental and Resource Sociology: Theoretical Issues and Opportunities for Synthesis." *Rural Sociology* 61(1):56–76.

Caldwell, Lynton K. 1983. "Environmental Studies: Discipline or Metadiscipline?" *Environmental Professional* 5(3–4):247–259.

Catton, William. 1982. *Overshoot: The Ecological Basis of Revolutionary Change.* Champaign, IL: University of Illinois Press.

CDC. No date. "Climate Effects on Health." Accessed November 1, 2016. http://www.cdc.gov/climateandhealth/effects/.

Clark, Susan G., Murray B. Rutherford, Matthew R. Auer, David N. Cherney, Richard L. Wallace, David J. Mattson, Douglas A. Clark, Lee Foote, Naomi Krogman, Peter Wilshusen, and Toddi Steelman. 2011. "College and University Environmental Programs as a Policy Problem (Part 1): Integrating Knowledge, Education, and Action for a Better World?" *Environmental Management* 47(5):701–715.

Conniff, Richard. 2012. "When Civilizations Collapse." Environment Yale blog. https://environment.yale.edu/envy/stories/when-civilizations-collapse/#gsc.tab=0.

Dirks, Arthur L. 1996. "Organization of Knowledge: The Emergence of Academic Specialty in America." Bridgewater, MA: University of Massachusetts. Boston. http://webhost.bridgew.edu/adirks/ald/papers/orgknow.htm. Retrieved January 2015.

Dunlap, Riley. 2013. American Sociological Association Environment and Technology section listserv, January 25.

Fischer, Joern, Robert Dyball, Ioan Fazey, Catherine Gross, Stephen Dovers, Paul R. Ehrlich, Robert J. Brulle, Carleton Christensen, and Richard J. Borden. 2012. "Human Behavior and Sustainability." *Frontiers in Ecology & the Environment* 10(3):153–160.

Foster, John B. 2012. American Sociological Association Environment and Technology section listserv, November 17.

Gaff, Jerry G., and James L. Ratcliff. 1997. *Handbook of the Undergraduate Curriculum: A Comprehensive Guide to Purposes, Structures, Practices, and Change*. San Francisco, CA: Jossey-Bass.

Hansen, James, Makiko Sato, Gary Russell, and Kharecha Pushker. 2013. "Climate Sensitivity, Sea Level and Atmospheric Carbon Dioxide." *Philosophical Transactions of the Royal Society A: Mathematical, Physical & Engineering Sciences* 371(2001):2–2.

Heinlein, Robert. 1952. "Where to?" *Galaxy Science Fiction* 3(5):13–22.

GBP website. Accessed November 1, 2016. http://www.igbp.net/about.4.6285fa5a12be4b403968000417.html.

Jackson, Wes. 2013. Public conversation with Wendell Berry and David Orr. Oberlin College, September 10.

Jacobs, Jerry. 2013. *In Defense of Disciplines: Interdisciplinarity and Specialization in the Research University*. Chicago: University of Chicago Press.

Klein, Julie T. 2005. *Humanities, Culture, and Interdisciplinarity: The Changing American Academy*. Albany: State University of New York Press.

Kolbert, Elizabeth. 2005. "The Climate of Man, Parts I, II, and III." *New Yorker*, April 25, May 2, and May 9 issues.

Lasswell, Harold D. 1971. "From Fragmentation to Configuration." *Policy Sciences* 2(4):439–446.

Lyall, Catherine, Ann Bruce, Wendy Marsden, and Laura Meagher. 2013. "The Role of Funding Agencies in Creating Interdisciplinary Knowledge." *Science & Public Policy* 40(1):62–71.

Mabe, Michael. 2003. "The Growth and Number of Journals." *UKSG Serials* 16(2):191–197.

Maniates, Michael. 2013. "Teaching for Turbulence." Pp. 255–268 in *The State of the World 2013: Is Sustainability Still Possible?*, edited by Worldwatch Institute. Washington: Island Press.

Miller, Asher, and Rob Hopkins. 2013. "Climate After Growth: Why Environmentalists Must Embrace Post-Growth Economics and Community Resilience." Santa Rosa, CA and Totnes, Devon UK: Post-Carbon Institute and Transition Network.

Mock, Brentin. 2015. "Want to Fix the Climate? First, We Have to Change Everything." Grist. April 2. http://grist.org/politics/want-to-fix-the-climate-first-we-have-to-change-everything/.

Moran, Emilio. 2010. *Environmental Social Science: Human Environment Interactions and Sustainability*. Malden, MA: Wiley-Blackwell.

Moran, Joe. 2010. *Interdisciplinarity*. London and New York: Routledge.

National Academy of Sciences. 2005. *Facilitating Interdisciplinary Research*. Washington, DC: The National Academies Press.

O'Brien, Karen. 2015. Great Transition Online Network Discussion, March 23. https://www.greattransition.org/contributor/karen-obrien.

Proctor, James D., Susan G. Clark, Kimberly K. Smith, and Richard L. Wallace. 2013. "A Manifesto for Theory in Environmental Studies and Sciences." *Journal of Environmental Studies and Sciences* 3:331–337.

Raskin, Paul. 2014. "A Great Transition? Where We Stand." Keynote address at International Society for Ecological Economics conference. http://www.greattransition.org/publication/a-great-transition-where-we-stand.

Rich, Nathaniel. 2013. "Writing the End." *New York Times Book Review*, April 19. http://www.nytimes.com/2013/04/21/books/review/writing-the-end.html?_r=0.

Rockström, Johan. 2015. "Bounding the Planetary Future: Why We Need a Great Transition." Boston, MA: Tellus Institute.

Rylance, Rick. 2015. "Grant Giving: Global Funders to Focus on Interdisciplinarity." *Nature Comment*, September 16. http://www.nature.com/news/grant-giving-global-funders-to-focus-on-interdisciplinarity-1.18344.

Simon, Julian. 1981. *The Ultimate Resource*. Princeton, NJ: Princeton University Press.

Smith, Kerri. 2009. "The Wisdom of Crowds." *Nature Reports Climate Change*. http://www.nature.com/climate/2009/0908/full/climate.2009.73.html.

Stafford, Susan G. et al. 2010. "Now is the Time for Action: Transitions and Tipping Points in Complex Environmental Systems." *Environment* 52(1):40–45.

Steffen, Will, Wendy Broadgate, Lisa Deutsch, Owen Gaffney, and Cornelia Ludwig. 2015. "The Trajectory of the Anthropocene: The Great Acceleration." *The Anthropocene Review* 2:1–18.

Steinbruner, John D., Paul C. Stern, and Jo L. Husbands. 2013. *Climate and Social Stress: Implications for Security Analysis*. Washington, DC, US: National Academies Press.

Stern, N. H. 2006. "The Economics of Climate Change." *The Stern Review*. Cambridge, UK; New York: Cambridge University Press.

Stern, Paul. 2013. American Sociological Association Environment and Technology section listserv, January 25.

Stern, Paul C. 1993. "A Second Environmental Science: Human-Environment Interactions." *Science* 260:1897–1899.

Stock, Paul, and Rob J. F. Burton. 2011. "Defining Terms for Integrated (Multi-Inter-Trans-Disciplinary) Sustainability Research." *Sustainability* 3(8):1090–1113.

Swanson, Donald R. 2001. "ASIST Award of Merit Acceptance Speech on the Fragmentation of Knowledge, the Connection Explosion, and Assembling Other People's Ideas." *Bulletin of the American Society for Information Science and Technology* February/March:12–14.

Tahir, Tariq. 2009. "Science Alone Will Not Save Us." *The Guardian*, May 19. http://www.theguardian.com/education/2009/may/19/social-sciences-tackling-climate-change.

Tainter, Joseph A. 1988. *The Collapse of Complex Societies.* Cambridge: Cambridge University Press.

Tenopir, Carol, and Donald King. 2014. "The Growth of Journals Publishing." Pp. 159–178 in *The Future of the Academic Journal*, edited by B. Cope and A. Phillips. Oxford: UK: Chandos Publishing.

Urry, John. 2009. "Sociology and Climate Change." *Sociological Review* 57:84–100.

Vesey, Laurence. 1965. *The Emergence of the American University*. Chicago and London: University of Chicago Press.

Victor, David. 2015. "Climate Change: Embed the Social Sciences in Climate Policy." *Nature* 520(7545):27–29.

Wei, Cynthia A., William R. Burnside, and Judy P. Che-Castaldo. 2015. "Teaching Socio-Environmental Synthesis with the Case Studies Approach." *Journal of Environmental Studies and Sciences* 5:42–49.

White, Damian. 2012. American Sociological Association Environment and Technology section listserv. November 18.

WHO website. "Climate Change and Human Health: Risks and Responses." Accessed November 1, 2016. http://www.who.int/globalchange/climate/summary/en/.

Zax, David. 2009. "The Last Experiment." *Seed Magazine*. http://seedmagazine.com/content/article/the_last_experiment/.

CHAPTER 2

Moving Toward Synthesis

In Greek, *syn* means together and *tithenai* means to put or place. Jointly, they give us the word synthesis, referring to the process or product of putting parts together, integrating them to create a whole. We can generally think of synthesis as the opposite of the more frequently used term, analysis. Though many today use the word pretty casually to refer to any old way of examining a thing, its literal meaning is rather specific. Analysis (also of Greek origin, *ana* means up, throughout and *lysis* means a loosening) refers to the process of trying to understand a thing by breaking it up into its constituent parts. Western science has been especially skillful at using this approach to attain all sorts of knowledge about how the world works. As we now know, however, there can be too much of a good thing. Like the king's army in the Humpty Dumpty nursery rhyme, we are struggling to put it all back together again.

In the context of this book, the demand for synthesis refers to pleas for the integration of that which analysis and academic specialization have taken apart, with the goal of forming a more complete, accurate, and useful picture of reality. The incongruity between the vastness of what we know and our apparent incapacity to use it to address socio-environmental problems makes the crisis of knowledge and the need for synthesis especially acute today. Interestingly, though, this recognition is not new. In fact, the demand for synthesis is centuries old.

D. Kasper, *Beyond the Knowledge Crisis*,
Palgrave Studies on Norbert Elias,
https://doi.org/10.1007/978-3-030-48370-8_2

This chapter explores the history of that demand and some noteworthy attempts to meet it and highlights key moments of tension between the dominant trend toward analysis and synthesizing countertrends in scientific and social thought. We see that, though we've been gradually inching our way toward a more integrated stock of knowledge, progress has been unsteady. A closer look at the general arc over time reveals the undulations of a series of smaller trends and countertrends. Consequently, the story below—the barest outline of the rich history of this trajectory—also does not proceed in a linear fashion. Instead, it jumps back and forth, tracing the movements toward and away from synthesis. Regardless of the status of these efforts, situating the demand for synthesis in a broader historical context—highlighting the obstacles, opportunities, and pitfalls encountered in past efforts—offers valuable lessons about what's getting in the way today and what we can do about it.

Wanted: Synthesist

There is increasing recognition that "society's most critical environmental problems are rooted in the multifaceted and deeply interconnected relationships between humans and the natural ecosystems in which they live" as the Socio-Environmental Synthesis Center explained on an earlier version of its website. As a result, there is a growing sense that we need "an integrated systems approach to synthesize data from diverse fields into a whole-systems perspective" (Stafford et al. 2010:41). Integration here does not mean blending things together to produce some kind of knowledge smoothie. Rather, it refers to the connection and coordination of highly differentiated "parts" within a coherent whole which results in maximizing the overall functionality and value of both. The parts in question here are the concepts, knowledge, theories, and methodologies relevant to socio-environmental systems at multiple levels.

History shows that this process is not inevitable. To the contrary, the synthesis of knowledge sought will require concerted efforts to create and help others envision, meaningful wholes from the pieces and parts revealed through analysis. Heinlein's wonderful term, "synthesist," provides a fitting title for this role, but one might be left wondering what exactly such work might entail. Here's a start at crafting a job description.

WANTED: SYNTHESIST

Qualified candidates should be adept at: recognizing complementary information and knowledge from diverse disciplines, integrating it into a unified framework, and using it to guide inquiry, facilitate communication and collaboration, and inform action on real problems. Requirements: long-term systems thinking, willingness to flout disciplinary conventions, ability to work well with others. Comes with exceptional benefits!

Though fictional and fairly general, this want ad points to some of the important functions a synthesist would serve and offers a way to identify those who have played this role, in some fashion, in the past. We know from Chapter 1 that, when it comes to socio-environmental synthesis, the position remains unfilled. What most don't realize is that the demand itself is not unique to our age. Spotting it in the historical record is not as easy as reading the want ads, but the call for synthesists nonetheless comes across loud and clear over the past three centuries, as the following examples demonstrate.

Take Auguste Comte, for instance. Those who have heard of the French philosopher are likely to have encountered him in an introductory sociology textbook, where he usually merits a passing reference for coining the term *sociology*—a Latin-Greek hybrid name for the then new science of society. Any coverage of him beyond that is likely to involve superficial mentions of his eccentricities and more radical ideas, allowing modern readers to dismiss him as a loon. But reading his work, as Norbert Elias (1978) suggests, with a sympathy informed by historical and personal contexts, one finds much to appreciate in Comte's 1830s writings, a great deal of which still rings true.

Like so many of our contemporaries, Comte was concerned with the specialization of knowledge. Admitting that many of the great advances made in modern times can be credited to the division of scientific labor, he urged readers to also acknowledge its disadvantages, including the ways that specialization confines the mind. When "almost everyone is busy about his own particular section," Comte observed, the result is the inability to embrace any one science in its entirety, much less the greater whole of scientific knowledge (1855:31). To arrest this harmful trend, without returning to an earlier undivided state of knowledge, Comte proposed that this division of labor must be perfected by "carrying it one degree higher—in constituting one more specialty from the study of scientific generalities" (1855:32). In other words, Comte was advocating for the role of the synthesist.

As he envisioned it, the business of synthesists would be to examine the physical, biological, and social sciences; ascertain their relations and mutual connections; and identify the fundamental principles which applied throughout. In turn, non-synthesists would be educated in a way which prepared them to appreciate the larger scope of science and be able to situate their particular discoveries within it. Ideally, advancements made on both sides would allow them to mutually correct and enhance each other's results, thus contributing to ongoing scientific progress.

At the time, Comte believed the sciences were well on their way to this arrangement and that the needed synthesis was imminent. With the benefit of hindsight, we know things didn't work out that way, or at least didn't move as quickly as expected. Long after Comte's petition, would-be synthesists would have reason to continue worrying about the fragmented state of scientific knowledge.

In the early part of the twentieth century, for example, South African statesman and philosopher Jan Smuts observed the futility of amassing more knowledge without a way to organize it. In a somewhat desperate tone, he urges readers to see that:

> It will not help merely to accumulate details...already becoming more than any individual mind can bear. New co-ordinations are required, new syntheses which will sum up and explain and illuminate the otherwise amorphous masses of material. (1926:6–7)

By the mid-1900s, those coordinations and syntheses had still not materialized. Although many celebrated the fantastic growth of knowledge about the universe, life, mind, and societies, they also recognized, as British biologist Julian Huxley did, that:

> large chunks of this new knowledge are lying around unused, not worked up or integrated into fruitful concepts and principles, not brought into relevance to human life and its problems. (1964:72–73)

And throughout the twentieth century, as awareness of poverty, disease, human conflict, and the threat of nuclear war grew, so did a general angst about humanity's capacity to effectively use hard-won knowledge to *do something* about these problems. At the close of the millennium, an uneasy blend of alarm and confidence continued to grace the pages of books about the state of human knowledge. American biologist, E. O. Wilson, represents this sentiment well, lamenting that:

> we are drowning in information, while starving for wisdom. The world henceforth will be run by synthesizers, people able to put together the right information at the right time, think critically about it, and make important choices wisely. (1998:294)

Yet, well into the twenty-first century, the US National Science Foundation concludes that our knowledge is still failing to keep pace with the challenges we face, which are so vast in scale and with such far-reaching consequences that they require "a different kind of scientific and social attention" (Stafford et al. 2010). Specifically, they argue, we need increased investment in research that strengthens our understanding of the links between processes in non-human nature and human activities, improving our ability to live sustainably on earth. In spite of growing agreement that "a unified science is needed soon and we have not even a common framework to unite these approaches," we seem to be making little progress. And the need becomes ever more pressing "as new problems of hideous complexity seem to arise almost yearly" (Waring and Richerson 2011:302 and 310).

This apparently poor track record might lead some to conclude that the synthesis dreamed of is just plain impossible. I'd like to suggest otherwise. Even if an all-encompassing synthesis of knowledge hasn't materialized as rapidly as desired, a closer look at the history of particular sciences reveals that there actually *has* been significant progress—in ways once thought impossible.

A Matter of Life and Matter

People today generally think of biology as a highly successful science and few question its worth. Groundbreaking work contributing to our understanding of disease-causing organisms, genetics, evolution, and more has earned biology a great deal of attention and respect. But this wasn't always the case. In the middle of the nineteenth century, the science of biology was viewed as very much behind the times compared to physics and chemistry. "Rather than a true science, it was a compendium of botanical and zoological curiosities" (Martínez and Arsuaga 2004:47). While it provided a great deal of information about the flora and fauna found in different places around the world, biology was in no way explanatory, was not advancing conceptually, and was thought to have reached a dead end.

One problem was that, with no big theory to hold it all together, it lacked a reliable means of organizing the information amassed over the years. This was due, in part, to an even more fundamental problem: scientists did not understand whether, much less how, the phenomenon of life relates to physical matter. Though perhaps difficult to imagine now, it was then typical among even renowned scientific thinkers to view life as something altogether alien to the stuff of the world. In the west, at least, the linkage between the two was considered an enduring mystery in scientific thought.

Two and a half millennia ago, Aristotle wrote about the *elan vital* believed to animate living beings. And still in 1840, it was perfectly acceptable for scientists to declare the notion of life to be "apprehended as a peculiar Idea, not resolvable into any other Ideas, such…as Matter and motion" (Whewell 1840:3). This apparently unbridgeable gulf between physical and biological phenomena also separated the sciences that studied them. To reconcile the perceived gaps between knowledge about matter (the realm of physics and chemistry) and emerging knowledge about life (the domain of biology) was the seemingly impossible problem of the day. Until that was resolved, there could be little genuine advancement in the life sciences, and no hope of integrating the sciences into a larger whole.

With the benefit of hindsight, we know that biology *did* make significant progress and today sits comfortably in and among the "natural sciences." One of the fundamental conditions making that possible was the development of a sound theory of the mechanisms of biological evolution. Among the many far-reaching effects of evolutionary theory, one was that it established a basis for understanding that life and mind as we know them emerge *from* matter. As such, they "are no longer aliens in a physical universe, but are…developments in and from the physical order" (Smuts 1926:10).

Although the particular mechanisms would continue to be debated, the basic notion of evolution via natural selection provided a framework for organizing what was known about the endless manifestations of life around the world. Not only was life now viewed as part of the physical order, but it could be understood as an ongoing process emerging in relation to the dynamic physical phenomena in which it is embedded, rather than as a substance about which scientists could only speculate. The implications were huge. The clarity that evolutionary theory brought to the enterprise of biology was so significant that Huxley declared Darwin to be "the 'Newton' of biology" (1964:9) and American biologist Theodosius Dobzhansky (1973) could legitimately proclaim that "nothing in biology makes sense except in the light of evolution."

Even in the context of revolutionizing ideas, though, concepts and thought patterns are not spontaneously transformed. Social thought tends to develop and change only gradually via the interplay of trends and countertrends of varying strength. The persistent controversy surrounding evolution in the United States illustrates this well.

Still, paradigms were shifting in the scientific world and, by around 1930, the tides in biology had noticeably turned. Although belief in an external vital force animating living beings hung on here and there, it was no longer a credible scientific position to hold (Woodger 1929). In purging ideas about the separateness of matter and life, biology solved what was thought to be an impossible riddle, renewing scientific enthusiasm for unlocking the secrets of the phenomenal world. But in its movement away from thinking of the difference between non-living and living entities as a matter of *substance*, biology was pulled in two seemingly opposite directions: reductionism and holism.

Looking to Smaller Parts

In the simplest terms, reductionism in science can be defined as an approach to understanding and describing complex phenomena in terms of the simpler phenomena and smaller parts that comprise them. Philosophers of science take this further, distinguishing between several different types of reductionism according to varying degrees of extremity, but we don't have to get that technical to appreciate that there are "stronger" and "weaker" forms of reductionism. We'll look at two of the forms most relevant here—epistemological, referring to *what* we can know, and methodological, referring to *how* we can know it—and some of their implications for science.

Epistemological reductionism, the more extreme form, is the idea that knowledge in one scientific domain can be reduced to another more fundamental body of scientific knowledge (Brigandt and Love 2008). In biology, this translates into the idea that it is possible, at least in principle, to "explain *all* biology in terms of physics and chemistry" (Crick 1966:10). Among representatives of this brand of reductionism was German-American biologist Jacques Loeb, who considered living organisms to be nothing but chemical machines from which life and consciousness appear as mere epiphenomena (Hammond 2003:35). In this view, such ephemera, because unobservable through classical physics-style analysis, have no place in science and must therefore be ignored.

These ideas famously influenced Loeb's student John Watson, who went on to apply this logic in psychology and to become the "father of behaviorism." Adamant about the need "either to give up psychology or else to make it a natural science" which progressed like medicine, chemistry, and physics, Watson's ultimate goal was to "control man's reactions as physical scientists want to control and manipulate other natural phenomena" (2009:6 and 11). The assumption that all phenomena can be explained in terms of the physical sciences led to the unfortunate neglect of questions about *which* concepts and techniques were appropriate to the subject matter and which were not. As a result, Watson and his fellow behaviorists cut their problems to fit their method, rather than the other way around. In their quest to study only the measurable and controllable aspects of behavior, they ended up studying mostly rats and pigeons and then extrapolating the findings to humans. The consequence of this behavioral revolution, American psychologist Harry Harlow observed, was the tendency "to focus more and more attention on problems of less and less importance" (1953:27).

Acknowledging that investigations of simple behaviors can be very informative about even simpler behaviors, Harlow continued, "very seldom are they informative about behavior of greater complexity" (1953:28). With far less diplomacy, biologist Ludwig von Bertalanffy echoed this sentiment, describing a large part of modern psychology as "sterile and pompous scholasticism which…covers the triviality of its results and ideas with a preposterous language bearing no resemblance either to normal English or normal scientific theory" (1967:6). The problem, as journalist and author Arthur Koestler (1978) put it, was not so much with viewing organisms as physical-chemical machines, but with viewing them as *nothing but* that. Taken literally, he says, "nothing butism" means that people could be described as "consisting of nothing but 90 per cent water and 10 per cent minerals—a statement which is no doubt true, but not very helpful" (1978:26).

Objections to epistemological reductionism, then, were not so much ideological as they were pragmatic. The approach just wasn't all that useful for understanding the complex phenomena of interest to many researchers. Put in contemporary biological terms, epidemiology may be related to molecular biology, which in turn is related to chemistry and ultimately to physics, but one can't effectively investigate a cholera epidemic "at the level of a molecule of cholera toxin or the quantum state of an electron around a single carbon atom within the toxin B subunit" (Fang and Casadevall 2011). The overall point is that, while efforts to analyze complex phenomena into their constituent elements are perfectly legitimate, and in fact indispensable, one must remember that a great deal is lost in the process.

Misgivings about attempts to explain complicated things directly in terms of their smallest parts generated resistance to reductionism, and fed misguided fear of a strawman, argues biologist Richard Dawkins (1996). Although some hard-core reductionists really did believe that the physics and chemistry of the day were sufficient to explain the behavior of living systems, most working biologists held the more moderate position that:

> the processes of living matter are subject to the same laws which govern the processes of dead matter, but that the laws operate in a more complicated medium; thus living things differ from dead things in degree and not in kind, and are, as it were, *extrapolations* from the inorganic. (Needham 1929:247)

The nature of these extrapolations and how to study them is the focus of the second form of reductionism we'll look at here.

Methodological reductionism, often traced back to Bacon and Descartes, refers to the idea that complex systems or phenomena can be studied and understood through analysis of their simpler components. Because this approach is more about what works in practice, and less about beliefs regarding ultimate knowledge, it is the less extreme form of reductionism. Notorious reductionist though he is, Dawkins recognizes the need for this more modest approach:

> if you try to use the laws of physics, in a naïve way, to understand the behavior of a *whole* living body, you will find that you don't get very far…it is doubtless true that at bottom the behaviour of a motor car is to be explained in terms of interactions between fundamental particles. But it is much more useful to explain it in terms of interactions between pistons, cylinders, and sparking plugs. (1996:10–12).

Dawkins uses the term "hierarchical reductionism" to describe the process of explaining complicated things in terms of interactions between their component parts, considered as successive layers of an orderly hierarchy. For any given level of complex organization, he explains, one may only attain a satisfying explanation by moving down the hierarchy one or two levels, *but not more*.

Wanting more—that is, trying to take the scientific explanation of a thing down to the most fundamental physical level—is what philosopher Daniel Dennett calls "greedy reductionism." The conflation of this with hierarchical reductionism is, some say, part of what has given reductionism (and reductionists) a bad name. As another well-known reductionist, E. O. Wilson (1998), explains, at each level of organization there are phenomena requiring new laws and principles which, at this point, still cannot be predicted from those at more fundamental levels. Wilson concedes that this kind of prediction may forever remain impossible, but (perhaps a little "greedily"), he thinks it's fun to try.

Regardless of how far we may someday get in explaining highly complex phenomena at the most fundamental level of physics, it is undeniable that methodological, or hierarchical, reductionism has been wildly successful, enabling scientists to gain insights into countless difficult and highly consequential questions. With it, they can ascertain, for example, the mechanisms of photosynthesis, why a virus fails to respond to drug therapy, and how vision works. At the same time, increasing attention to the limitations of reductionism has engendered somewhat of a backlash, as with the recent accusations of excessive reductionism in molecular biology (Fang and Casadevall 2011). Within a larger historical context, this contemporary pushback can be viewed as part of counter-efforts toward holism.

Parts, Wholes, and Relational Patterns

Jan Smuts is usually credited with coining the term holism in his 1926 book *Holism and Evolution*. In addition to his concerns with disciplinary fragmentation as an obstacle to knowledge development and problem-solving, Smuts worried that the study of life, once thought incompatible with physical science, had gone too far in the other direction, becoming a mere province of it. The concept of holism in biology emerged in opposition to both mechanism, which sees living beings as complex machines, and vitalism, which holds that life comes from an external animating force. It represents the view that biological processes can only be adequately explained if organisms are viewed, not as substances or entities unto themselves, but as particular patterns of organizing relations.

Thus far, the notion of holism in biology has evoked mixed reactions. Depending on how it's conceptualized, it has "either found general recognition in biology or been dismissed as incompatible with the biological facts" (Mittelstrass 2014:51). Part of what makes it dismissible are the connotations (especially in medicine) that the word presently carries of "alternative" and "unscientific" (Freeman 2005). But these associations should not deter one from appreciating Smuts' original intent in introducing the term. His aim was to highlight nature's tendency toward the evolution of wholes, a tendency beautifully described by biologist Tyler Volk:

> Layer by layer, entities were constructed or evolved from the previous layer of wholes. These previous wholes became parts for the next layer, and so on, in a glorious effusion of stratified beings. It's a spiral process: Parts turn into wholes, which turn into parts for still more encompassing wholes. (1998:48)

One of the implications of this realization, Koestler (1978) noted, is that wholes and parts, in any pure sense, don't really exist in living systems. What can be seen as a whole at one level can be viewed as a part of some superseding whole at another level. The liver, for example, can be seen as a whole at the level of organs, but a part in the context of the digestive and excretory systems. The same can be said of the cells, tissues, organs, and systems of any organism, and of organisms themselves in the context of larger ecosystems. Koestler proposed the word *holon* as an alternative concept to capture the hybrid nature of units in complex living systems and to facilitate synthesis-oriented thinking. *Holarchy* is the term he used to describe a hierarchy of somewhat self-regulating holons. Although the terminology never quite caught on, thinking in terms of larger patterns of organizing relations did.

From early on, more holistically-minded biologists insisted on viewing an organism as "a certain type of biological organization," rather than a mere collection of parts (Woodger 1929:296). With an eye to the whole, these organicists, as they were then called, came to recognize that the relational patterns biologists studied exhibited a distinctly higher level of complexity and qualitatively different laws than those observed at the level of non-living phenomena. Biologists in this camp viewed the discovery of these laws as the chief task of their discipline; tremendous optimism surrounded this new direction. There was impassioned confidence, for example, that it would bring about "a fundamental change in the world picture" (von Bertalanffy 1972:410) and that the "hierarchy of relationships, from the molecular structures of the carbon compounds…to the equilibrium between species in ecological wholes…will probably be the guiding idea of the future" (Needham 1932:92).

At the same time, it was recognized that, "though the need for a comprehensive biological science is great, the difficulties in obtaining it are equally considerable" (Needham 1936:6). Likewise, some understood that, while the general ideas were easy to understand intuitively, the challenge was "to make these notions precise in order to enable us to see how we can use them for scientific purposes" (Woodger 1930:8). They were right. Despite earnest efforts, the concept of holism in science remained counter to the more dominant reductionist trend. As Koestler (1978) observed, it went against the scientific *zeitgeist* of the time.

But even if the spirit of the time was *apparently* reductionist, there is growing consensus that, in reality, it was never a matter of one versus the other. Some scientists now argue that the notion of methodological reductionism versus holism is a false dichotomy and that they are not really in opposition at all (De Backer et al. 2010). In biology, for example, "reductionistic and holistic methodological approaches have been coexisting and thriving for centuries" (Fang and Casadevall 2011). Overall, there is increasing recognition that either approach by itself is too limiting, that they should correctly be viewed as interdependent and complementary, and that only in applying them together can scientists make real progress (Kitano 2002; Mazzocchi 2012; Rose 1997; Schuster 2007). Rather than any sort of pure study of parts and wholes, the thinking goes, we need more of a two-way street, where we can move in either direction as need dictates. Even if this approach hasn't yet permeated scientific methodology, it has nonetheless cropped up here and there, exerting important influences and leaving valuable insights in its wake.

A Two-Way Street

Over the course of the twentieth century, a number of scholars proposed ways to make this complementary approach more explicit and to eradicate the either/or spirit that seemed to carry the day. Upon observing the lack of well-established concepts for these different but corresponding operations, sociologist Norbert Elias, for example, suggested the terms "analytical" and "synoptic" (1956:244-246). In an analytical approach, he explains, the theoretical representation of a system or whole is treated mostly as the background against which questions about the constituent parts stand out as the main object of study. A synoptic approach, on the other hand, involves steps to form a more coherent theoretical representation of a system or whole as a unifying framework and potential testing-ground for theories about relationships among the constituent parts.

To explain mutual relationships of influence within systems, philosopher and social scientist Donald Campbell (1974) introduced the terms "upward" and "downward" causation. Later, biologist and systems theorist Tyler Volk (1998) coined the phrases "outward influence" and "inward influence." The former refers to the specific ways that parts (in relation to other parts) create the conditions that bring a given whole into being, and the latter refers to how wholes effect changes in the parts that comprise them. Outward influence, Volk explains, can be seen in the evolutionary order of parts/wholes. After all, molecules could only have come into existence after the atoms that make them up; in the same way, cells could come into being only after molecules, and multi-cellular organisms after cells. This logic is apparent in non-living systems too. The development of a car, for instance, was necessarily preceded by the invention of parts like wheels and axels, gears and brakes, pistons and connecting rods, and so on. But outward influence isn't only evident in the *evolution* of a thing. We can also see it in the ways that parts regularly influence the functioning of wholes. Think, for instance, of how a malfunctioning part, like a heart valve or a flat tire, affects the whole.

Inward influence moves in the other direction, with the whole influencing the parts. Even though cells, for example, preceded multi-cellular organisms, whole organisms have also served as the milieu within which more and more types of cells have developed. The whole machine we call a car has similarly functioned as a basis for the development of numerous variations on the same basic theme (Volk 1998:48-50).

Whatever the terminology, Elias argues, working with "problems whose framework represents a highly integrated unit depends in the long run on the co-ordination and balance between steps in both directions" (Elias 1956:245). The main takeaway here is that, in order to understand complex phenomena, we need to be able to shuttle back and forth between the different levels of the parts and wholes (or holons and holarchies) in question. Whether we imagine the movement to be up and down or in and out is less important than *that* we move in both directions.

E. O. Wilson agrees, explaining that moving "top down across two or three levels of organization at a time by analysis, then bottom up across the same levels by synthesis" is the two-step procedure by which natural scientists should work (1998:74). The ultimate goal is the assembly of knowledge through the linking of facts and fact-based theories across disciplines. Today, many recognize the dire need for an effective integration of knowledge for understanding and addressing socio-environmental problems. We also know that, despite a great deal of progress in this direction, we're not there yet. Moments of triumph in the more recent past, however, led some to believe that they *had* reached, or at least were very close to, a point at which humans knew enough to finally take control of their destiny.

This genre of efforts at synthesis in the name of human progress failed. It's important to understand why, lest we assume that their project was doomed by some inherent flaw in the aspiration itself. The reasons for their failure can be boiled down to premature confidence, inadequate conceptual and organizational structures, and garden variety interpersonal conflict. Seeing how these played out in the following situations offers a sense of how *not* to go about this work while also providing valuable insights into the substantial changes in disciplinary structures which advancements toward synthesis will likely require.

In the Name of Progress

With insight into the mechanisms of biological evolution, the hubris that came with increasing abilities to manipulate nature, and a new appreciation for the uniqueness of humans' capacity for cultural adaptation, some scientists came to view humanity's presence on the global scene as representing "a new *state* of evolution, a new *phase* of the cosmic process" (Huxley 1964:264). These conditions, along with technological developments which made warfare a luxury the human species could no longer afford, revived arguments that it is "necessary for us to evolve the institutions requisite for social life in this new world" (Goldschmidt 1959:14). This marked a new phase in the perceived "need to transcend disciplinary boundaries and emphasize unification over specialization" (Morawski 1986:219).

This time the challenge would be to integrate the social into the biophysical sciences. The resulting comprehensive and coherent framework, it was dreamed, would integrate:

> the scattered and largely unutilized resources of our knowledge, and order them to provide a new vision of human destiny, illuminating its every aspect, from the broad and enduring sweep of cosmic process to present-day polities, from the planetary web of world ecology to the individual lives entangled in it, from the dim roots of man's past to the dawning possibilities of his far future. (Huxley 1964:74)

Despite echoes of the old Enlightenment dream, things were a bit different. New elements in social thought—especially evolutionary thinking, ideas about the unconscious, and discoveries of inherent uncertainties in the physical world—challenged traditional notions of inevitable and linear progress, the preeminence of rationality, and our predictive powers. In this new context, intellectuals were called to task "to assert the feasibility of comprehending that buzzing confusion of human actions" (Morawski 1986:221).

So began a new phase of efforts toward a synthesis of knowledge, supported enthusiastically by wealthy foundations who would fund high profile projects at some of the nation's most prestigious institutions. As it turns out, confidence in these endeavors was premature. Their failings nevertheless provide insights relevant to the development of a viable socio-environmental synthesis, and so are worth a closer look.

The Institute of Human Relations

"The time has come once again," proclaimed Yale President James Angell in 1929, "to attempt a fruitful synthesis of knowledge." He was referring to the Institute of Human Relations (IHR), formed the year before with funding from the Rockefeller Foundation. Its general goal was "improving humankind," which it sought to do so through breaking down barriers in science and integrating all research "that pertained to the study of man" (May 1971:141). The kinds of applications envisioned for this unified human science mostly fell under a loosely defined human engineering agenda (Bryson 2015). Though broadly conceived (involving medicine, law, biology, sociology, and other disciplines), psychology was at the center of this program and its quest to understand the dynamics of human action.

In terms of conventional measures of productivity, the IHR initially appeared successful, but with respect to the goal of scientific synthesis, it didn't fare so well. The new thinking inspired by Darwin and Freud turned out to be trickier to deal with than anticipated. Although most of the scientists studying humans acknowledged the multi-causal, interdependent, and irrational nature of human action, the concepts and paradigms at their disposal could not accommodate this new outlook. The resulting perplexities led IHR directors to restructure the organization in order to impose greater order. Trying to redesign the weekly seminars in ways that supported the development "of a social science with 'power' like physics," they invited physical scientists who were deemed to have the right mental mechanisms and exhorted participants to read Newton's *Principia* in order to acquaint themselves with those mental abilities (Morawski 1986:233). In the end, the IHR would rejuvenate the older intellectual worldview.

In a decidedly reductionist spirit, researchers returned to tacit notions that the human sciences must ultimately be anchored in a prime unit. Having chosen individual behavior as the correct unit, "the move…to behaviorism, especially for psychologists, was relatively uncomplicated, and many economists, political scientists, and sociologists endorsed behaviorism as well" (Morawski 1986:221). Ultimately, their work did not yield the kind of unified theory they were after, nor the human improvements they envisioned. By the end of the IHR's first decade, the original ideal of an integrated human science had been replaced by a search for universal and mechanical laws of individual behavior.

The Rockefeller Foundation became increasingly critical of the results reported and funding gradually decreased, ending altogether in 1949. A subsequent Ford Foundation grant enabled the IHR to hire five more researchers, from five different disciplines, tasked anew with developing a theoretical synthesis. Proceeding much as their predecessors had, they ended up reproducing similar kinds of output. With no integrative theoretical framework, these efforts too fell apart and the IHR was dissolved in 1963.

The Department of Social Relations

Harvard professors attempted a similar synthesis, albeit with just the social sciences. In 1942, on the heels of decades of inter- and intra-departmental conflict, a handful of faculty began to petition for the formation of a new department—the Department of Social Relations (DSR)—combining sociology, clinical psychology, cultural anthropology, and social psychology. They argued that the DSR put Harvard at the forefront of changes that will eventually be forced in other institutions "because of the excessive diversity of subject matter" (Allport 1944) and that rigid departmental boundaries did not reflect the nascent synthesis forming (Allport and Boring 1946). Eventually, circumstances tipped the scales in their favor and, in the summer of 1946, the Department of Sociology was dissolved, and the Department of Social Relations was formed and granted authority to award degrees. The DSR attracted record numbers of graduate student applications and the undergraduate program grew rapidly, becoming the second largest undergraduate concentration at the university by 1950 (Johnston 1998). There were problems almost immediately.

The DSR offered no courses providing models of integrative scholarship, nor did it instruct students on how to synthesize different kinds of subject matter. As quickly as it grew, the program developed a reputation for easy courses and began to attract the school's weaker students, making meaningful integration that much more difficult. Also, because at the graduate level, advanced degrees were only available in individual disciplines, not in Social Relations, the dissertation remained the only hope for graduate students to demonstrate explicitly interdisciplinary work. The absence of reliable training to do so, however, lowered the probability of success. In terms of faculty, the fact that the DSR, as George Homans described it, was "a department built on hatreds" didn't help matters (Johnston 1998). Unsurprisingly, the formation of this program did not prevent new interpersonal conflicts from arising. And, because most appointments were made on the basis of disciplinary reputation, in many ways the DSR reinforced specialization.

Like Yale's Institute of Human Relations, the DSR produced numerous publications—500 in its first decade alone—but there was little evidence of interdisciplinary activity. Sixty percent of the publications were single-authored and a substantial number of others were co-authored by members of the same discipline (Schmidt 1978). Research was, at best, multi-disciplinary and "fragmentation, not integration, prevailed" (Johnston 1998:36). The most systematic attempt at theoretical integration was the collaborative effort, supported by the Carnegie Foundation, resulting in the 1951 volume, *Toward a General Theory of Action*, edited by Talcott Parsons and Edward Shils. Parsons thought of it as "his baby" and something of a manifesto for the enterprise of synthesis that the DSR represented (Smelser 2001:vii). Although Parsons felt this collaborative venture had been a success, others did not agree. Though widely read, it was deemed "an important failure" in social science theory (Beals 1953:422). Reviewers tended to be generous, acknowledging the monumental task of unifying multiple fields, but in the end remained mostly unpersuaded, as did other members of the DSR. Their rejection of Parsons' urging that the book be "adopted as the official doctrine of the department," at one infamous meeting, ended the DSR's official effort of theoretical integration (Johnston 1986). Intensifying internal stresses, along with changing disciplinary trends nationwide, induced the sociologists in DSR to go their own way in 1970. What remained became the Department of Psychology and Social Relations until 1986, when faculty voted to change the name to Department of Psychology, erasing "Social Relations" from Harvard University.

The Center for Advanced Study in the Behavioral Sciences

Having to distribute a considerable sum of money after Henry Ford's death in 1947, Ford Foundation trustees commissioned a report which, reflecting the atmosphere of postwar America, laid out a five-point program to support activities which would contribute to: world peace and a world order of law and justice, allegiance to freedom and democracy, economic well-being of all people, improved educational opportunities, and increased knowledge of the factors which influence and determine human behavior. The Center for Advanced Study in the Behavioral Sciences (CASBS), established in 1952, grew out of this fifth program. Motivated especially by concerns about destructive technologies and increasingly polarized international relations, and dissatisfied with disciplinary fragmentation, CASBS was designed to bring together scholars in the behavioral sciences. The CASBS's founders viewed interdisciplinary research as absolutely necessary to solving humanity's problems and their work was touted as the social science version of the Manhattan Project (Hammond 2003).

The CASBS never did produce a landmark synthesis theory of human behavior, but unlike the others, it still exists. Its website mentions the original charge CASBS was founded to serve—"to increase knowledge of factors which influence or determine human conduct, and extend such knowledge for the maximum benefit of individuals and society"—but appears to have dropped the goal of integrating them. There are opportunities for "sustained conversation" and "exposure" to new methods, perspectives, theories, and problems, but the 40 or so fellows CASBS currently sponsors annually seem to work mostly on individual projects within their disciplines, rather than explicitly collaborative work.

As an interesting sidenote, the early phases of this project gave rise to another synthesis-oriented effort. Shortly after CASBS was founded, four of the original fellows discovered a common interest in systems theory and established the Society for General Systems Research in 1956. Although nobody "had much ambition to construct a general theory of everything," President Kenneth Boulding (1985) wrote, "the feeling that the whole universe must be in some sense a single system was not wholly absent."

The goal was to develop a framework that would enable specialists to understand one another. That goal and the organization (renamed the International Society for the Systems Sciences, ISSS, in 1988) survive today, but at the time of this writing remain somewhat counter to more powerful specializing trends. As incoming president of the ISSS, Debora Hammond, reminded annual conference attendants: "the original goal of the organization was to foster the unity of science," but there is enormous resistance from faculty, among whom there is still serious "confusion about what is meant by unity of knowledge" (Hammond 2005).

While the organization's more than 60 years is worth celebrating, it's membership numbers, relative to individual disciplines, give one pause. The organization's vice president informed me in an email in 2016 that paid membership in ISSS stood at approximately 300, with about 1500 on the mailing list. When compared to paid membership statistics of the primary social science disciplinary organizations (which in 2016 were at 13,000 for the American Political Science Association, 14,000 for the American Sociological Association, 18,000 for the American Economic Association, and 137,000 for the American Psychological Association), we begin to see how marginal the movement for synthesis still is. This should not come as a surprise. If the excitement for interdisciplinarity synthesis is to be productive, Kenneth Boulding points out, "it must operate within a certain framework of coherence" (1956:200). And the fact is, we simply don't have one.

Recognizing Possibilities, Realizing Potential

In the absence of a coherent framework, some say, we've moved in the opposite direction, with the tide turning in the 1960s toward a more "reductionist form of Darwinism across a number of biological disciplines, and the premature specialization and fragmentation of the social sciences" (Quilley and Loyal 2005:828). To complicate matters further, a good deal of work done by sociologists during the latter part of the twentieth century was rooted in a general distrust of *any* general theory. They confined themselves increasingly to investigating isolated clusters of problems which could be explored using methods established within certain natural sciences, "though they themselves lacked what these others already possessed: a more unified, more highly integrated system of theoretical constructs as a common frame of reference for isolated studies of part-connections" (Elias 1956:245–225).

This brief glimpse at the history of efforts to move toward synthesis usefully reveals three major obstacles: a disproportionately reductive approach, the absence of a coherent theoretical framework, and attachment to disciplinary identities and structures. These have not only hindered the establishment of more integrated knowledge, but they have also exacerbated specialization and its problems. Are we to conclude from all this that integration of the social and natural sciences is a lost cause? Or does it only seem that way, much like prospects for integrating the physical and biological sciences once did?

I take the latter position and contend that a meaningful synthesis of knowledge is not only possible, but within our reach. Achieving it, though, will require fundamental structural changes—especially in the social sciences. This is the focus of the next chapter, which examines the status of the social sciences, some of the challenges affecting the overall trajectory of a science of human social life, what can be done to correct course, and why it is so very important that we do.

References

Allport, Gordon. 1944. In the Allport Papers, July 24. Harvard University Archives.

Allport, Gordon, and Edwin Boring. 1946. "Psychology and Social Relations at Harvard University." *American Psychologist* 1(4):119–122.

Beals, Ralph. 1953. "Review of *Toward a General Theory of Action*," edited by Talcott Parsons and Edward Shils. *American Anthropologist* 55(3):422–423.

Boulding, Kenneth E. 1985. "The Boulding's Eye View," unpublished paper in Kenneth Boulding Collection, University of Colorado.

Boulding, Kenneth E. 1956. "General Systems Theory—The Skeleton of Science." *Management Science* 2(3): 197–208.

Brigandt, I., and A. Love. 2008. "Reductionism in Biology." In *The Stanford Encyclopedia of Philosophy*, edited by E. N. Zalta. http://plato.stanford.edu/archives/fall2008/entries/reduction-biology/.

Bryson, Dennis. 2015. "Mark A. May: Scientific Administrator, Human Engineer." *History of the Human Sciences*, 1–35.

Campbell, Donald. 1974. "Downward Causation in Hierarchically Organised Biological Systems." Pp. 179–186 in *Studies in the Philosophy of Biology: Reduction and Related Problems*, edited by Francisco Jose Ayala and Theodosius Dobzhansky. London: Macmillan.

Comte, Auguste. 1855. *The Positive Philosophy of Auguste Comte*, translated by Harriet Martineau. New York: Calvin Blanchard.

Crick, F. 1966. *Of Molecules and Men*. Seattle, WA: University of Washington Press.

De Backer, P., D. De Waele, and L. Van Speybroeck. 2010. "Ins and Outs of Systems Biology vis-a-vis Molecular Biology: Continuation or Clear Cut?" *Acta Biotheor* 58:15–49 [PubMed].

Dawkins, Richard. 1996. *The Blind Watchmaker*. New York and London: W. W. Norton.

Dobzhansky, Theodosius. 1973. "Nothing in Biology Makes Sense Except in the Light of Evolution." *The American Biology Teacher* 35(3):125–129.

Elias, Norbert. 1956. "Problems of Involvement and Detachment." *The British Journal of Sociology* 7(3): 226–252.

Elias, Norbert. 1978 [1970]. *What Is Sociology?*, translated by Stephen Mennell and Grace Morrissey. New York, NY: Columbia University Press.

Fang, Ferric, and Arturo Casadevall. 2011. "Reductionistic and Holistic Science." *Infect Immun* 79(4):1401–1404. Published online February 14. https://doi.org/10.1128/iai.01343-10.
Freeman, Joshua. 2005. "Towards a Definition of Holism." *British Journal of General Practice* 55(511):154–155.
Goldschmidt, Walter. 1959. *Man's Way: A Preface to the Understanding of Human Society*. New York: Holt.
Hammond, Debora. 2003. *The Science of Synthesis*. Boulder: University Press of Colorado.
Hammond, Deborah. 2005. "Whither Systems?" Presidential Address to the International Society for the Systems Sciences. http://isss.org/bulletins/2006_GeneralSystemsBulletin_v35_no_member_list.pdf.
Harlow, Harry. 1953. "Mice, Monkeys, Men, and Motives." *Psychological Review* 60(1): 23–32.
Huxley, Julian. 1964. *Essays of a Humanist*. New York: Harper & Row.
Johnston, Barry. 1998. "The Contemporary Crisis and the Social Relations Department at Harvard: A Case Study in Hegemony and Disintegration." *American Sociologist* Fall:26–42.
Johnston, Barry. 1986. Personal interview with George Homans. March 11, William James Hall, Harvard University.
Kitano, Hiroaki. 2002. "Systems Biology: A Brief Overview." *Science* 295:1662–1664.
Koestler, Arthur. 1978. *Janus: A Summing Up*. New York: Random House.
Martínez, I., and J. L. Arsuaga, J. L. 2004. *Green Fire: The Life Force, from the Atom to the Mind*. New York: Thunder's Mouth Press.
May, Mark. 1971. "A Retrospective View of the Institute of Human Relations at Yale." *Behavior Science Notes* 6:141–172.
Mazzocchi, Fulvio. 2012. "Complexity and the Reductionism–Holism Debate in Systems Biology." *Interdisciplinary Reviews: Systems Biology and Medicine* 4(5):413–427.
Mittelstrass, Jürgen. 2014. "Complexity, Reductionism, and Holism in Science and Philosophy of Science." *Acta* 22: 45–53.
Morawski, J. G. 1986. "Organizing Knowledge and Behavior at Yale's Institute of Human Relations Author." *Isis* 77(2): 219–242.
Needham, Joseph. 1936. *Order and Life*. New Haven: Yale University Press.
Needham, Joseph. 1932. "Thoughts on the Problem of Biological Organization." *Scientia* 52:84–92.
Needham, Joseph. 1929. *The Sceptical Biologist*. London: Chatto and Windus.
Quilley, Stephen, and Steven Loyal. 2005. "Eliasian Sociology as a 'Central Theory' for the Human Sciences." *Current Sociology* 53(5):809–830.
Rose, Steven. 1997. *Lifelines: Biology, Freedom, Determinism*. New York: Penguin Press.

Schmidt, Patrick. 1978. *Towards a History of the Department of Social Relations*. Harvard University.

Schuster, Peter. 2007. "A Beginning of the End of the Holism Versus Reductionism Debate?" *Complexity* 13(1):10–13.

Smelser, Neil. 2001. "Introduction." Pp. vii–xix in *Toward a General Theory of Action*, edited by Talcott Parsons and Edward Shils. New Brunswick, NY: Transaction Publishers.

Smuts, Jan. 1926. *Holism and Evolution*. Westport, CN: Greenwood Press.

Stafford, Susan G., et al. 2010. "Now Is the Time for Action: Transitions and Tipping Points in Complex Environmental Systems." *Environment* 52(1):40–45.

Volk, Tyler. 1998. *Gaia's Body: Toward a Physiology of Earth*. New York: Copernicus.

von Bertalanffy, Ludwig. 1972. "The History and Status of General Systems Theory." *Academy of Management Journal* 15(4):407–426.

von Bertalanffy, Ludwig. 1967. *Robots, Men, and Minds: Psychology in the Modern World*. New York: George Braziller, Inc.

Waring, Timothy M., and Peter J. Richerson. 2011. "Towards Unification of the Socio-Ecological Sciences: The Value of Coupled Models." *Geografiska Annaler Series B: Human Geography* 93(4):301–314.

Watson, John B. 2009 [1924]. *Behaviorism*. New Brunswick and London: Transaction Publishers.

Whewell, William. 1840. *The Philosophy of the Inductive Sciences London*, Vol. II. London: John W. Parker, West Strand.

Wilson, E. O. 1998. *Consilience: The Unity of Knowledge*. New York: Vintage Books.

Woodger, Joseph. H. 1930. "The 'Concept of Organism' and the Relation Between Embryology and Genetics Part I." *The Quarterly Review of Biology* 5(1):1–22.

Woodger, Joseph. H. 1929. *Biological Principles: A Critical Study*. New York: Routledge.

CHAPTER 3

A Science of Human Social Life? Present State, Future Prospects

Despite fervent attempts in the early to mid-twentieth century to develop, integrate, and apply the social sciences to some of the world's most serious problems, things did not progress as expected. Humanity's capacity for destruction grew, inequalities widened, prejudice and violence persisted, and environmental harms intensified. There were improvements too, of course, especially in the form of expanded educational opportunities, equal rights, and information technology, but warriors of Progress discovered that humanity's trajectory was not as amenable to scientific control as they once assumed. In contrast with earlier expectations that a science of social life would guide the reorganization of society and lead humanity ever closer to perfection, few today are *that* optimistic about its potential. Whatever value social science is granted is conceived of in much narrower terms.

Does this reining in of expectations reflect the ultimate futility of trying to use the social sciences for the good of humanity and the world? If so, how can we more effectively use what we've got? If not, and a more effectual social science *is* possible, how can we achieve it? These questions are more important now than ever. When all evidence points to the need for massive societal transformation in the context of a growing recognition that sustaining human life on the planet "requires a social avalanche of unprecedented proportions" (Fischer et al. 2012), there is renewed

D. Kasper, *Beyond the Knowledge Crisis*,
Palgrave Studies on Norbert Elias,
https://doi.org/10.1007/978-3-030-48370-8_3

interest in figuring out how to better use the social sciences to get from "here" to "there."

To date, however, there remains little clarity about *how* to do that, which fuels skepticism about the utility of the social sciences. In response, social scientists proclaim the worth of their endeavors more loudly, and the cycle of attack and defend repeats. In order to exit this pattern, we need a mindset of curiosity rather than defensiveness in examining basic questions like: What is social science? What is its subject matter? What fundamental premises is it based on? How does it relate to other sciences? How can it best inform the questions that now beset us and what methods are most appropriate? We should also remember that there was a time when the natural sciences were in a similar position.

For the vast majority of human history, people mostly attributed disease, disaster, and death to forces beyond nature—which they could try to appease but never control. Even as recently as a few centuries ago, few would have turned to scientists for help in dealing with these sorts of problems. To the contrary, it once seemed absurd to believe that humans could directly intervene in any meaningful way. Reconstructing this former state of not knowing expands our view and offers a helpful analogy to contemporary skepticism about the social sciences, prompting us to ask what happened in the natural sciences to change the situation? Answers to that question can help us imagine the possibility of comparable advancements in social science. This chapter explores that possibility.

It begins by reviewing the development of ideas about why social science is necessary, what constitutes the social sciences, and to what degree they have been successful, or not. The focus then shifts to look more specifically at sociology, originally viewed as *the* overarching social science. In confronting over a century of critiques of sociology we get much needed clarity about what sociology is, what it's for, and how a more coherent sociology can not only bring about advancements in the social sciences, but also in socio-environmental studies.

Where Do the Social Sciences Stand Today?

One way to begin to answer this question is to compare social science's present situation with original expectations. For Comte, writing around the time of the advent of formal social science disciplines, the only legitimate way to study social phenomena was to view "each element in the light of the whole system" (1855:462). Though he recognized this would be no easy matter, he insisted that a synthetic sense of social science as a whole must precede any specialization, lest the whole enterprise be poorly carried out. As we know, even though a coherent overall system of the social sciences has not yet been achieved, extensive specialization *has* occurred. Let's take a look at some important moments in that process, beginning with efforts to determine what social science is.

Defining Social Science

Divisions in the social sciences were beginning to crystallize in the first half of the nineteenth century, but it wasn't until after 1850 that they began to reflect today's familiar disciplinary structures. National and historical contexts determined which subject matters rose to the top of what was coming to be called "social science," initially including history, economics, sociology, and political science, with geography, psychology, and law as close seconds. The list of disciplines constituting the social sciences was gradually institutionalized through the establishment of university chairs and departments offering disciplinary degrees, accompanied by disciplinary journals, associations, and library collections (Wallerstein 1996).

By 1945, economics, political science, and sociology had emerged as the three primary social sciences—differentiated from the rest by their nomothetic character, that is, their quest to find general laws governing human behavior. Notable overlaps among them soon prompted people in those fields to "stake out their separate terrains as essentially different" (Wallerstein 1996:31). Almost as quickly as they were established, though, changes in world politics, the growing scale and intensity of human activities, and the expansion of the university system raised new questions about what belonged under the heading of social science. Over the years, a variety of answers were generated, none of them definitive.

In 1996, the US Congress created the Office of Behavioral and Social Sciences Research (OBSSR), with an initial mandate to establish a standard definition of research in those areas. "Heretofore, there had been no single definition of the field that could be used to assess and monitor" funding support (National Institutes of Health). The resulting "definition" didn't mention specific disciplines, but proposed categories of research with dozens of subsets of criteria. A decade later, the National Academies would publish their own taxonomy of the social sciences. To the three original cores of economics, political science, and sociology, they added psychology, communication, geography, linguistics, and public policy, with several sub-areas under each, plus three "emerging areas": criminology, science and technology studies, and urban studies (National Academies 2006). In the meantime, OBSSR's definition continued to be periodically revised (Riley 2016).

While uncertainty about what counts as social science persists, two things have become very clear. Specialization has been powerfully at work within and between what have variously been considered social science disciplines, and there remains no reliable way of organizing social sciences and their subspecialties (unless you count alphabetical order). So far, history seems to confirm Comte's hunch. Without a general framework to guide our study of the particulars, we end up with specialized discussions which only exacerbate the crisis of knowledge and negative perceptions of the social sciences.

In stark contrast to initial expectations that social scientific knowledge would "lay down a definite basis for the reorganization of society" (Comte 1848:2), enabling us to mitigate conflict, promote prosperity, and deliberately shape humanity's future, there is a widespread sense today that the social sciences are of little practical value. With some exceptions, they are generally viewed as "soft"—less rigorous, exact, objective, and therefore less important than, the "hard" sciences.

Perceptions of Social Science

Not being part of standard American high school curricula, social science disciplines are unfamiliar to many, making it difficult to accurately gauge public attitudes toward them. Unfavorable assessments, on the other hand, are vividly apparent among those in positions to impact the social sciences through decisions about funding. In 2012, for example, the US House of Representatives passed an amendment to eliminate National Science Foundation (NSF) funding for political science research. Some believed the amendment didn't go far enough, declaring that "the NSF shouldn't fund any social science," in part because, despite the use of quantitative methods in political science, economics, and psychology, "these disciplines can never achieve the objectivity of the natural sciences" (Lane 2012). The following year, during negotiations for a spending bill, a change was proposed to prevent the NSF from "wasting federal resources on political science projects, unless the NSF Director certifies projects are vital to national security or the economic interests of the country" (Mole 2013).

The US House of Representatives went for broader change in 2015, introducing a bill which would cut funding for the "Social, Behavioral, and Economics Directorate," a branch of the National Science Foundation, by 45%. Even though this branch comprised only 5% of the entire NSF budget, the cut was deemed symbolically important. Strong proponents of such cuts argued that funding for social science should not come at the expense of other sciences which are more likely to "produce breakthroughs that will save lives, create jobs, and promote economic growth" (Matthews 2014). While more moderate supporters of the cut acknowledge that social science has some value, they argue that it just isn't as important as other kinds of "real" science that give us new medicines, enhance national security, and support economic growth.

This sentiment is not unique to the United States. In Europe, where prospects for the social sciences have generally been more favorable, social science disciplines are also on the defense. When a new discipline-blind funding program was being launched to promote the integration of the social sciences and humanities into all research, for example, European Commission President Jean-Claude Juncker diverted part of that budget into an investment fund aimed at boosting Europe's economy (Rabesandratana 2014). In 2015, more than two dozen Japanese universities announced plans to close or scale back programs in the social sciences and humanities in the effort to promote what the Prime Minister Shinzo Abe described as "more practical vocational education that better anticipates the needs of society" (Grove 2015). In 2019, Brazil's president announced a plan to defund sociology programs in order to shift financial support to "areas that give immediate returns to taxpayers, such as veterinary science, engineering, and medicine" (Bolsonaro 2019).

Aspiring to Relevance

While American social scientists boldly assert the value of their work, they have been forced to reckon with these public relations problems. In the aftermath of real and attempted funding cuts, the American Political Science Association has devoted significant resources to convincing the public of its worth. Its 2014 report, for example, focused on "Improving Public Perceptions of Political Science's Value," which they sought to do primarily through updating their means of disseminating research. The following year's report, "Let's Be Heard! How to Better Communicate Political Science's Public Value," explored ways to cater to the preferences of online information consumers.

Motivated by similar desires, in 2009 the American Psychological Association created its first ever strategic plan and began implementation in 2011. In addition to typical organizational action items (e.g., assessing workforce demand, expanding funding support), one of its chief goals was to increase recognition of psychology as a science. The overall purpose of these diverse initiatives, the organization's CEO Norman B. Anderson stated, is "in a word, *relevance*" (APA 2011:S4).

In the 1980s and 1990s, panic in sociology escalated with the closing (and near-closing) of departments at prominent US universities (ASA 1978, 1989; Pearman 1992). At the same time, declining numbers of majors prompted a flurry of re-framing and marketing activity (Das 1989; Fabianic 1991). Since then, the American Sociological Association (ASA) has devoted much of its annual reporting to measuring sociology's "success," mostly in terms of numbers of majors, post-graduation job attainment, student satisfaction, and graduate degrees, and has more recently turned its attention to gauging sociology's reach through social media.

Certainly, efforts to enhance visibility, communication, and perceptions of relevance are good things. The problem, however, is that none of the above strategies feature any critical assessment of the disciplines themselves or their approaches to social science research. The inordinate emphasis on growth—in numbers of students, subscribers, online traffic, and funding dollars—is indicative of the presumption that things are fine, and that they need only amplify disciplinary voices, expand resources, and garner more respect from the public. Without more objective self-reflection, it's difficult to discern what the real problems are.

Engaging in this kind of reflection requires a loosening of emotional involvements with hopes, fears, and ego- (and discipline-) serving ideas—a capacity Elias termed "detachment." He thought of this capacity as being on a continuum, with "involvement" at the opposite pole, along which the tendencies of people and groups can move in either direction. The increasing capacity for detachment in the natural sciences, he argued, especially when it "gained the upper hand in human knowledge of non-human nature," was a crucial factor in their rapid advancement (Elias 1987:xxix). It is likewise necessary for meaningful progress in the social sciences. This is especially important in sociology, viewed by some as poised to serve as *the* organizing structure for the social sciences, though not in its current form (Carroll 2013; Kasper 2016; Quilley 2010; Quilley and Loyal 2005).

Helping us understand why sociology is not up to the task are critiques from within the discipline itself. These minority voices highlight ongoing tensions between dominant trends and lesser-known countertrends (not unlike the movement between synthesis and analysis discussed in Chapter 2) and thus offer insights into the development of sociology's trajectory to date. Importantly, they go beyond critique, also providing clues for what it takes to establish a sound sociology. To begin, we turn to one of the most frequently asked questions.

What Happened to Sociology?

At the midpoint of the last century, "it appeared that sociology would take its place at the table of science by providing scientific explanations for the operative dynamics of the social universe," Jonathan Turner explained in a presidential address to a regional association (2008:281). That the dream was never realized prompted him to wonder, "what happened to sociology?" Although this sort of question doesn't come up much in polite sociological company, there has been a small but steady stream of critics who have felt compelled to inquire.

Some have bluntly asked "what's wrong with sociology?" (Cole 1994, 2001; Davis 1994). Others have focused on specific concerns about sociology's theoretical incoherence, the absence of agreed upon core concepts, and a perceived lack of relevance and practical value—without which, they conclude, sociology is not a sustainable enterprise (some of the many possible examples include: Becker and Rau 1992; Carroll 2013; Cole 1994; Davis 1994; Deflem 2013; Goldthorpe 2000; Huber 1995; Kalberg 2004; Kasper 2011; Keith 2000; Michalski 2008; Phillips 1999; Rule 1994; Stinchcombe 1994; Turner and Turner 1990; Turner 2006, 2008).

Though persistent, this line of questioning has been mostly background noise within sociology. The few responses it has elicited tend toward a defense of theoretical pluralism as a sign of a healthy discipline. This justification, however, doesn't satisfy critics who point out that genuine pluralism, involving a vigorous competition of ideas, requires consensus about the fundamentals, which sociology does not have. As they see it, sociology's pseudo-pluralism cannot offer any resolution to its "general, and steadily worsening, disarray" (Goldthorpe 2000:7). Without concerted effort to assess and address sociology's problems, not only did the issues not go away, they got bad enough to cause some prominent sociological figures to lose heart.

Peter Berger, for example, threatened to revoke his once passionate *Invitation to Sociology*, "so as not to be responsible for yet more innocent students being seduced into what may well be a bankrupt enterprise" (Berger 1992:12). Irving Horowitz, in *The Decomposition of Sociology*, laments the discipline's squandered potential. Tracing this disappointment back further to 1959, we find C. Wright Mills coming to terms with sociology's failure to deliver on its promise in his book, *An Autopsy of Social Science*. He re-titled it *The Sociological Imagination* to reflect a hope that perhaps the social sciences were not dead after all, but merely sick (Geary 2009). Fifty years earlier still, we discover Albion Small, influential founder of American sociology, complaining that, for all the attempts to develop a science of society, "there is very little to show" except for "object-lessons in how not to do it" (Small 1905:43).

From the point of view of sociology's contemporary critics, not much appears to have changed. Mathieu Deflem argues that, despite the present illusion of success, sociology has been in perpetual crisis, remains unable to deliver on its original promise, and "is not generally thought to be relevant to society" (2013:162). Though perhaps not easy for sociologists to hear, these observations have been instrumental for clarifying what is wrong with our discipline. It is now crucial to understand why. Underlying sociology's theoretical incoherence and perceived irrelevance is a basic confusion about what sociology is and what it's for.

What Is Sociology?

As a college student, I used to dread the inevitable question, as predictable as talk about the weather: *what's your major?* Upon answering "sociology," people would either ask "what's sociology?" or, equating it with psychology or social work, they would proceed to comment on my choice. I didn't have a good response in either case and assumed the problem was my own ineloquence. Eventually, I began to wonder why I didn't have a better answer.

Because sociology is not taught in most high schools, perhaps this confusion is understandable (DeCesare 2007, 2005). However, one might reasonably expect professional sociologists to have a clearer sense of things. A closer look at their own answers suggests otherwise. Introductory textbooks, intended to acquaint people with the discipline, reflect inconsistencies about what sociology studies and what its core concepts are (Kasper 2011; Keith and Ender 2004). Professional associations don't add much clarity. The ASA's definition is a catchall that includes: the study of society, the social lives of people, groups, and societies; our behavior as social beings, social aggregations, and "an overarching unification of all studies of humankind, including history, psychology, and economics."[1] The British Sociological Association (BSA), echoing that last point, describes sociology as "the one social science which embraces the whole range of human activities."[2]

The problem is not that these characterizations are wrong, as we'll see, but that they don't lend themselves to a clear sense of what sociology actually is and how it differs from other social sciences. Without this it's difficult to see, for instance, what subspecialties like economic sociology, political sociology, and social psychology contribute to these areas that economists, political scientists, and psychologists don't already cover. This lack of clarity only reinforces accusations, observed long ago, that sociology has no subject matter of its own and that sociologists' continued "failure to agree upon a definition of their science, or upon precise description of their task" only confirms this impression (Small 1905:4).

At the root of this problem is the fact that one of sociology's most fundamental concepts, *society*, remains ill-defined. The observation that "conceptions of society are as varied as the conceptions of sociology" is as true today as when it was made well over a century ago (Ellwood 1907:304). Out of convenience, most equate society with "nation" (Bauman 2002; Elias 1978, 1991a; Mann 1986), but it has also long been used to refer to cultures, geographic regions, or groups (Ellwood 1907). Some avoid the problem altogether by simply not defining it. In one best-selling textbook, for example, sociology is defined as "the study of human society," but nowhere in its more than 700 pages is "society" defined (Conley 2019:4). Consequently, the term remains, as ever, "a veritable will-o'-the-wisp," a thing impossible to catch and see (Small 1905:184). In frustration, one sociologist suggested abolishing the concept altogether (Mann 1986:2). In any case, the failure to develop a reliable understanding of society has profoundly affected the development of sociology. In particular, it has hindered our ability to explain the relationship between individual and society, preserving its status as *the* driving question of contemporary sociology (Calhoun et al. 2007:4; Elliott 1999:7; Elliott and Ray 2003: xiii–xiv; Ritzer and Goodman 2008:500) and keeping sociology stuck in a conceptual cul-de-sac.

At the turn of the twentieth century, some plausibly argued that sociology's confusion was due to its newness as a science. A definition, it was argued, "can only be formulated when the final stages of the development of the science have been reached" (Ellwood 1907:300). Well into the twenty-first century, sociology is no longer new yet the confusion endures. The main difference is that today not many seem to know or care. Conversations about the fundamentals tend to happen at the margins, allowing sociologists to go about their specialized business as though all were well. Things could have gone differently, and still can. In order to appreciate the alternatives possible, we must address the question of why we need sociology at all.

What Is Sociology for?

As originally conceived, sociology's chief purpose was to provide a scientific theory of how social change happens. Whatever differing ideas there were about the uses to which such a theory could be put, understanding social change was thought to be one of sociology's main tasks. The fact that it was still unfinished in the late twentieth century was taken by some as evidence of its impossibility or undesirability, sparking a postmodern turn in sociology away from "objectivity" and toward a more local, event-based narrative with overt moral intent (Seidman 1991). In the early 2000s, these ideas manifested in the form of "public sociology."

Few took issue with public sociology's stated aim of bringing sociology into dialogue with audiences beyond the academy. Concerns were raised, however, about its stated promise to make sociology "a moral and political force" (Burawoy 2007:25). Who decides, critics wondered, what's moral and which values to pursue? Former ASA president Frances Fox Piven offered an answer, explaining that public sociology strives "to address the public and political problems of people at the lower end of the many hierarchies that define our society…to the marginalized and down-and-out" (2007:163). In addition to prompting questions about qualifying criteria for those categories, public sociology introduces other dilemmas.

Are we to take a stand, one sociologist asks, against the boycott of Israeli academics or do we support the Palestinians? Is a sociologist who expresses deeply felt religious beliefs against extending marital benefits to gay partnerships doing public sociology? Likewise for the sociologists who later oppose hiring him because of those statements (Smith-Lovin 2007:131). "Achieving an accurate understanding of the social world," Massey argues, "requires a willingness to admit that one's own views might be misguided or just plain wrong," but public sociology provides no criteria for determining (2007:146). As a result, critics say, this particular attempt to make sociology more relevant only worsens its problems.

Without a firm empirical foundation, sociologists are left standing on unsteady ideological and moral ground. As many have argued, the overtly ideological nature of public sociology has contributed to the perception of sociology as inherently leftist (Cole 2001; Deflem 2013)—a reputation which has caused "even more doors to be slammed in our faces" (Turner 2005:28) and further weakened the discipline's future prospects by attracting students interested in activism without academic rigor (Stinchcombe 2007; Deflem 2013). More troubling still is the lack of rigor observed in formal academic arenas, where evidence suggests that in some areas "[s]cholarship based less upon finding truth and more upon attending to social grievances has become firmly established, if not fully dominant" (Pluckrose et al. 2018).[3]

While critics of public sociology are passionate about applying sociological knowledge to real-world problems, for them, advocating for one side over another in moral and political battles is *not* what sociology is primarily for. Reflecting on his life's work in *What Use Is Sociology?*, Zygmunt Bauman offers a different answer to this question. The sociologist's job, he explains, is not to solve problems according to one's hopes and fears, nor to impose a certain set of values. Its real task is to identify and convey the complex networks we are embedded in, and in so doing, to purvey orientation in a changing world. Put differently, he says, the point of sociology is not to impose choices, but "to render the available choices intelligible and the responsibilities involved evident" (Bauman 2014:71).

This view echoes the recurring conclusions of sociologists working counter to the mainstream. Though far less prominent, their arguments have been remarkably consistent over the years. Rather than being directly involved with political commitments, they contend, sociology can best "underwrite more effective political interventions" by providing a more effective means of orienting people within social relational dynamics over the long term (Quilley and Loyal 2005:817). With this, we will be better equipped to make the mostly unseen and uncontrolled processes shaping our lives "more accessible to human understanding" and thus to greater control (Elias 1978:153). It is only when people are able to realistically orient themselves in the world that they have the possibility of being able to more consciously shape their own history (Van Den Bergh and Van Benthem 1986).

We Need Sociology More Than Ever

Sociology emerged as radical changes in the wake of Europe's age of political, industrial, religious, and cultural revolutions sparked a desire to understand what was happening and why. The pace, scale, and effects of such changes have only increased since then, as have the size and complexity of the social networks in which we find ourselves embedded. It still rings true that "attempts to constitute a science of society would not have been so obstinate, nor pursued in ways so various, if an instinctive need of it had not been deeply felt" (Comte 1855:450).

The present "deepening interdependence binding humanity and Earth into a single community of fate" provides an even greater incentive for more accurately orienting ourselves so as to better see and control the consequences of our actions (Raskin 2016:ii). When, say, "driving a car down the New Jersey Turnpike and secreting a shell in the South Pacific—turn out to be connected" (Kolbert 2006:235), we need some help connecting the dots. The kind of sociology we need, one that helps us draw these linkages and thus orient us in the world, begins with a more expansive view, one that shows humans within their complex constellations of interdependent relations *and* places them within the biological and physical (henceforth biophysical) contexts they emerge from, act on, and which, in turn, act on them.

At a time when human activities are known to be altering the conditions for and thus the composition of life on this planet in ways which are making it less habitable, less secure, and less able to sustain the human enterprise as we know it, humanity is at a crossroads. Major changes are on the horizon no matter which way we go, and it's in our best interest to be able to guide them, or at least anticipate them so as to be able to navigate what lies ahead. This is the context in which socio-environmental researchers are seeking a reliable theory of social change and why devoted sociologists have come to view their discipline's chaotic status as tragic "at a time when the world needs sociological knowledge" (Turner 2006:27).

The good news, as the rest of this book shows, is that we already have what we need; we just have to put the pieces together in the right way. As with any construction project, this re-assembly will involve some creative destruction—mainly in the form of letting go of current notions about sociology and the social sciences in order to be able to re-imagine them as needed.

Re-imagining the Social Sciences

It can be difficult to let go of ideas, even when we know they're badly flawed, and especially when they've been heavily institutionalized. Precedents of successful re-visioning, however, assure us that change *is* possible. Remember biology's situation in the mid-nineteenth century? As we saw in Chapter 2, biology couldn't explain things as physics and chemistry could, was not advancing conceptually, and was thought to have reached an impasse. Its riddle was to explain life without reference to an external animating force. Until biologists began to see life as a question of *organization* rather than *substance*, they were stuck. Clearly, biology got itself unstuck. Today, its numerous subspecialties fit together within a coherent framework of "biological sciences," encompassing all the divisions of science "examining various aspects of vital processes."[4]

Sociology has long been stuck at an analogous impasse. It has amassed enormous quantities of information, but with no empirically-grounded means of organizing it all, relies, as biology once did, on amorphous ideas of external forces and remains incapable of providing a parsimonious or reliable explanation of human social processes. Time and again, sociology's counterthinkers pointed to shifts in perspective that would get sociology unstuck and moving in the right direction. In particular, they advocated viewing sociology's subject matter in terms of dynamic relations rather than things, recognizing the social realm as a distinct but inseparable level of phenomena, and allowing these insights to inform a more appropriate organizational structure for the social sciences. Although they never gained enough momentum to move the discipline as a whole in these directions, it's worth looking at the points where things could have gone differently, and still can.

Relational Processes

In the first decade of the nineteenth century, some were already clear about the fact that human society is not a thing, nor a species of things, but that "it is relationships between persons and the world they occupy…processes linked together into *n*thly compound processes" (Small 1905:184). In this view, progress in sociology meant shifting attention from "society" to "human association," which meant defining sociology as the study of patterns of reciprocal interactions among people and changes in them (Ellwood 1907). These arguments did not win the day and the idea of society as a thing repeatedly "seduced sociologists into profitless dialectics" (Small 1905:184). Fortunately, that didn't stop others from bringing back the message, again and again.

In the 1930s, British sociologist, Morris Ginsberg tried to persuade his fellow sociologists that their job is to study "the whole web or tissue of relationships into which human beings enter with one another" and their long-term patterns of growth and decay (Ginsberg 1937:166). Over the following five decades, Norbert Elias worked tirelessly to promote a relational, and thus "genuinely sociological," conception of society to counter the habit of substantialist thinking and of seeing individuals as self-contained entities and societies as a jumble of adjacent but unconnected spheres of activity. In the late twentieth century, observations that large segments of the sociological community still viewed the social world mostly in terms of static substances, rather than dynamic relations, inspired efforts to institute a more relational sociology. In "Manifesto for a Relational Sociology," Mustafa Emirbayer (1997) argued that the choice between substantialist and relational modes of inquiry represents a choice of foundational assumptions about the very nature of reality and a key dividing line in sociological investigation.

Plenty of others agreed. Relational sociology gained a strong following, just not strong enough. It currently remains a specialized pursuit in sociology, "largely unknown and generally misconstrued, if not totally rejected" (Donati 2011:25). A focus on collective and individual entities as units of analysis still dominates (Crossley 2011). Sociology's enduring failure to embrace the social—"the patterned connections or relationships between or among constituent elements"—as the core organizing feature of the discipline is viewed by contemporary critics as remarkably ironic (Michalski 2008:533).

Interestingly, this very shift—from thinking in terms of unchanging entities to dynamic relations—has been one of *the* defining features of major scientific advancements. Physics went from seeing the universe as static to viewing it as being in constant flux. Geological explanations for changes in the earth moved from catastrophism (the idea that landforms stayed the same until a sudden catastrophe came along to alter them) to plate tectonics and ongoing subterranean processes. Biology discarded ideas about immutable species and adopted a view of ongoing evolution. In psychology, the idea of fixed brains and homuncular minds gave way to an understanding of both brains and minds as processes of continual formation and re-formation. Genuine advancement in sociology requires the same sort of paradigm shift, the achievement of which demands that sociology finally takes physical and biological phenomena seriously.

A Distinct, but Inseparable, Level of Phenomena

For the most part, sociology appears to have ignored early arguments that "any competent theory of human associations...must square with knowledge about those physical and vital relationships upon which the later social phenomena rest" (Small 1905:420). Instead, sociologists have gone in the opposite direction, exhibiting a general aversion "to attempts to elucidate the relationship of the level of integration they themselves study to the preceding levels, and particularly to the biological" (Elias 1987:154). The results have not been good.

The problems cited above can largely be attributed to the futility of studying disembodied social phenomena. Recognition of this, along with the increasing need to address urgent socio-environmental problems, is part of what prompted the formal launch of "environmental sociology" in the 1970s. The goal was to situate societies firmly in their biophysical contexts and help rid sociology of its long-standing anthropocentrism (Catton and Dunlap 1978). As assessments in subsequent decades showed, however, sociology failed to assimilate this basic logic (Buttel 2002; Catton 1994; Foster 2002; Kasper 2016) and remains "steadfast in [its] quest for disembedded 'social' causal variables" (Carolan 2005:12). Without grounding in knowledge about the biophysical conditions that comprise human social phenomena, "sociology and disciplines within social science lack a firm ontological basis" (Carroll 2013:10), leaving their subject matter "without anchorage in the observable world...left hanging in the air" (Elias 1991b:43).

Just as biology made the shift to seeing life as emergent *from* the physical order, rather than alien to it, sociology needs to more explicitly acknowledge the biological and physical conditions from which human social life emerges and on which it depends. Doing this means abandoning both the anti-science attitude that has plagued sociology for decades (Black 2000; Huber 1995; Salter 1996, 2012) and the eagerness to emulate classical physics, an "exceptionally inappropriate" model for sociology (Lieberson and Lynn 2002:2). Though ostensibly opposite positions, they have produced similar obstacles for sociological thinking. In both cases, the antidote is to situate *the social* within a general hierarchy of phenomena, including the physical and biological.

This is not difficult. We can easily see that reality does not consist of a realm of physical phenomena in one place, biological happenings in another, and human stuff in a third cordoned off region. We don't experience them as separate, yet that's often how we think and talk about them, referring to "natural sciences" and "social sciences" as if they sit arbitrarily side by side. A hierarchical mental model would facilitate a more reality-congruent way of understanding them as interrelated, albeit distinct, levels of phenomena. Whatever uniqueness we might claim as humans, there is no denying the fact that human social life is comprised of people (biological organisms) who are themselves comprised of the matter and energy which make up all observable phenomena. Figure 3.1 depicts the basic idea.

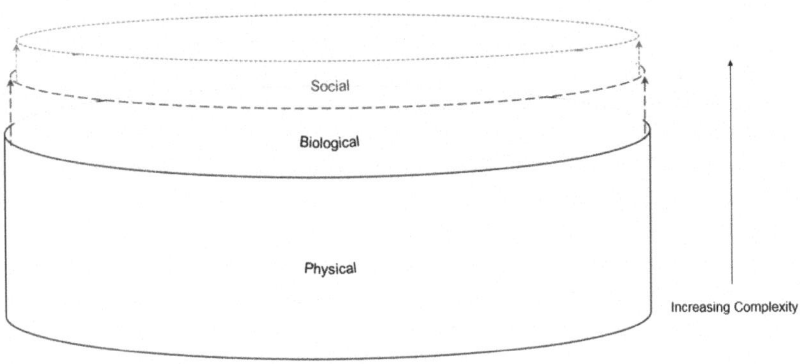

Fig. 3.1 Simple hierarchical model of fundamental categories of phenomena

In this way of thinking, all matter and energy are categorized as physical, but we can think of some things as *just* physical. Along this hierarchy, physical phenomena are most amenable to explanation via investigation of their part-units (e.g., particles, atoms, elements, molecules). Biological phenomena encompass the physical, but are different in ways that call for entirely new concepts, like *life* and *death*. Here, there is greater functional interdependence in and among the composite units of interest (e.g., cells, tissues, organs, systems, unicellular organisms, multi-cellular organisms), the properties and behaviors of which are more influenced by the organization and integration of their constituent parts. Comprised of biological and physical phenomena, social phenomena emerge from those levels and are subject to the laws at work in them—*but there's something more going on* which constitutes a distinctly different level with its own set of governing principles. The "something more" here is the layer of meaning that permeates and characterizes the experience of human social life. Here, in addition to the biological and physical processes taking place, there's a universe of symbols and their dynamic meanings at work. Elias called this the "fifth dimension" (1991b:47).

Wind and calcium, for example, are physical phenomena. In the category of more complex biological phenomena are the birds, insects, and plants who make use of the wind to fly and distribute seeds and the colonies of bacteria living on the calcium-containing teeth with which an organism chews food. All this *and more* is happening at the social level, where wind can be imbued with personality or symbolize the Divine, as with the Holy Spirit in Judeo-Christian traditions. Birds can be used to represent meekness, freedom, power, peace, death, and other qualities. Bacteria can be imagined as tiny invaders to be eradicated, beneficent aids to our internal biodiversity, or long strings of genetic codes. The state of the teeth in a human mouth can be used to interpret age, attractiveness, or socioeconomic status. On it goes. Here in the five-dimensional human realm, biophysical phenomena like birth, death, thunder, darkness, tea leaves, vocal sounds, and drawn shapes, just for starters, can be understood to *mean* something in a way that is not the case (as far as we can tell) at the levels of "mere" physical or biophysical phenomena. And *what* they mean can vary almost infinitely.

Increasing Complexity

The nested nature of phenomena is easy to intuit, but it takes a bit more work to understand what's behind the increasing complexity throughout the three levels. Some explain it in terms of quantities, arguing that things get more complex mainly because there are so many more variables and thus more potentialities at play in succeeding levels "up" the hierarchy. Clearly, quantities matter. As Stephen Hawking explains, quantum mechanics allows us, in principle, to predict nearly everything we see around us, within the limits set by the uncertainty principle. *In practice,* however, "we cannot solve the equations for any atom besides the simplest one, hydrogen, which has only one electron" (Hawking and Mlodinow 2005:100–101). To establish probabilities for more complex atoms physicists must rely on powerful computers and approximations.

This reality underscores the monstrous proportions of such a task applied to living cells and multi-cellular organisms! Ratcheting this basic logic up several notches, physicist Murray Gell-Mann famously challenged his colleagues to imagine how hard physics would be if particles could think. Differences in complexity, however, are not just a matter of quantities. As Elias (1956) explains, it's a matter of *qualitatively different stages of integration and differentiation in the relational patterns of organization* of parts and units across levels. A simple analogy can help us unpack what this means.

Take written English, a system made up of a series of units and their respective constituent parts. Lines and shapes form letters, letters combine to make words, configurations of words form sentences, and so on up through paragraphs, chapters, books, and genres. The number of lines and shapes is crucial in the formation and discernment of individual letters. Adding to or taking away parts of a letter produces something entirely different. But it's not just the *number* of parts that matters. The letter A, for example, involves three line segments, but if the middle line is moved to the bottom it ceases to be an A and becomes a triangle. At this stage of integration and differentiation (of and between lines and shapes), we can see that interpreting them as letters is a matter of how those parts are *organized in relation to one another.* The level of words represents a different stage of integration and differentiation, involving combinations of letters which form sound patterns, prefixes, suffixes, roots, and so on. Individual letters (and the lines and shapes that make them) remain fundamental to this level, but we don't focus on them to derive meaning.

Instead, we look to the patterns of relation and organization among letter combinations. Likewise, sentences derive their meaning from the relational patterns of organization among words, phrases, and punctuation—yet another stage of integration and differentiation. Moving up through the various levels, we notice different rules at work in each.

In similar fashion, viewing levels of the phenomenal world as a continuum of complexity allows us to more readily compare degrees of integration and differentiation in organizational patterns, and to recognize different kinds of "laws" at work at different stages in them. At one end of the continuum are units in which the constituent parts are more temporarily and loosely associated. They can exist independently of each other while retaining their characteristic properties. Think molecules, configurations of atoms which can assemble and disassemble repeatedly without being significantly changed. At the other end of the spectrum are units which operate more as open, dynamic, self-regulating systems. Characterized by a high degree of integration and interdependence among their constituent parts (which themselves consist of hierarchies of interlocking part-wholes), these *cannot* be isolated without radical changes in their properties or in the unit itself. Assembly and development processes at this end are characterized by greater irreversibility. While hydrogen and oxygen atoms can come together, separate, and be reunited again to form water molecules, for instance, an animal or plant cannot be disassembled to the point of death and then revived through reassembly.

The logic of this continuum, Elias (1987) contends, sheds light on the relevant degrees of difference between humans and non-humans and their respective social units. It also highlights the need to consider *which* methodological devices are appropriate at which level. Those useful in studying simpler units, for example, may lose their function or become merely auxiliary when applied to more complexly organized phenomena, just as handwriting analysis and sentence diagramming, though useful for certain purposes, are not all that helpful in discerning a book's themes. In general, the more that subjects under investigation exhibit characteristics of highly self-regulating, dynamic, interdependent systems, the more are system- and process-models required for exploring and conveying regularities among parts and wholes. This approach takes us beyond simplistic either/or ways of thinking in terms of reduction or holism, induction or deduction, and instead demands that we can nimbly move back, forth, and among models of the whole and those of its parts.

This idea of hierarchical phenomena is not at all new, but it remains oddly alien in sociology (and other social sciences), where references to human biology in discussions of human social phenomena range from sporadic to non-existent (Black 2000; Kasper 2014). In some ways, this is not surprising, Elias (1956) points out. History shows that those exploring new kinds of problems initially model their concepts and procedures after those which worked well in other fields. It takes time for a new group to emancipate itself from the dominant style of thinking. In the meantime, their attitude toward established ways is likely to oscillate between uncritical submission to the authority and prestige of the dominant standards or rejection and denial of the functions older models had or have in the development of newer ones (Elias 1956:239). From this pattern of tensions, broader changes eventually take shape.

While this perspective is illuminating, the urgency with which new insights are demanded of the social sciences makes it crucial that we do what we can to bring about results sooner rather than later. Because the results sought are impeded by the current structure of the social sciences, it has been argued, part of the remedy is a revised structure that better reflects reality, especially of humans in the biophysical world.

Re-structuring the Social Sciences

As Chapter 1 reminds us, university departmental structures are poor representations of the nature of reality. They tend to reflect a combination of tradition, local circumstance, and competition for resources, rather than a concerted effort to reflect the phenomenal world as it is. The institutionalization of the social sciences in particular, it's been noted, has resulted in "some major exclusions of reality" (Wallerstein 1996:95).

The tacit assumption that a discipline corresponds "to a certain segment of the empirical world" promotes an especially distorted view (Boulding 1956:197), encouraging an "extremely artificial image of society as a hotch-potch of adjacent but unconnected spheres" (Elias 1978:141–142). Most people don't actually believe social reality is carved up this way, but without more adequate conceptual tools, disciplinary structures unwittingly convey and reinforce a distorted picture. Recognizing how this works against our efforts to understand and potentially gain greater control over the direction of human social processes, some have lobbied for an all out revolution, demanding a social science that: takes history and the biophysical world seriously, can better handle complexity, and more accurately represents the interrelatedness of economic, political, cultural, and other forms of human social activity (Flyvbjerg 2001; OECD 2004; Wallerstein 1991; Wallerstein 1996).

Those who have ventured to imagine the necessary re-structuring tend to agree that it involves re-thinking sociology's position in the social sciences. Because the social sciences study different aspects of a "singular, emergent level of reality," it has been argued, we should think of sociology "not as a self-contained discipline but as a nexus" (Carroll 2013:4). As such, it is poised to provide an overarching theoretical framework and play "an orchestrating role in a broader human science, encompassing phenomena...engendered by distinct but intertwined biological and social planes of integration" (Quilley and Loyal 2005:808). This sentiment is well aligned with sociology's early advocates, who argued that sociology, understood correctly, refers not just to one fragment of social science, but to all investigations into "the process of human association from its minutest to its largest phenomena" (Small 1905:184).

To be sociological, then "is not the exclusive purview of persons called sociologists. It is an obligation of all social scientists" (Wallerstein 1996:98). A re-organization which establishes sociology as an umbrella for the scientific study of human social life writ large may seem radical to some and impossible to others (though it already aligns with definitions from the ASA and BSA). Given the attachment to disciplinary identities and competition for resources that tend to evoke a "circle the wagons" response to protect disciplinary turf at all costs, one might reasonably wonder whether the effort is worth it (Carroll 2013:17).

This is an important question, but history reminds us that there are moments when major realignments are called for. "The question before us is whether the early twenty-first century may be another such moment" (Wallerstein 1996:73). I believe it is. I also believe that Elias' approach—informed by biology, vigilant of the influence of emotional involvements, and accounting for relational processes, as it does—provides the makings of the sociology we need, one that naturally takes our biophysical contexts into account and thus to explain complex interdependent processes and the history they generate over time.

Whatever changes a re-structuring brings about (e.g., in titles and labels, organizational charts, identities, and research foci), they would be a small price to pay for advancing the twin causes of more effectively understanding socio-environmental processes and increasing our collective ability to effectively act on that understanding. Two initial steps can nudge us in the necessary direction.

The first involves labels and conceding that "perhaps sociology will have to be re-invented by another name" (Turner 2006:27). "If a new word, more encompassing than 'sociology,' has to be invented to signify the theoretical-empirical study of the broadest questions of human co-existence," others agree, "then so be it" (Gabriel and Mennell 2011:19). Biology offers an example of a simple and minimally disruptive solution. Just as the term "biological sciences" encompasses the various sciences that deal with phenomena at the biological level, the moniker "sociological sciences" could be used to refer to the various sciences that study phenomena at the social level. In this way, specialists of particular ***kinds*** of relational processes (e.g., political, media, family, economic, etc.) can continue to identify as such, but their specializations would fall under the overarching designation of the sociological sciences. But what about sociology, one might ask, would it just disappear, being subsumed by the sociological sciences? No, but its character would likely change. As within the biological sciences, where "general biology" is now considered a subdivision concerned broadly with the origins and life processes of living organisms, perhaps an analogously general form of sociology would specialize in long-term, large-scale trends of dynamic human social interdependence.

The second step involves properly situating the subject matter of the sociological sciences in the physical and biological conditions within which it occurs. This would produce a general sociological theory at a high level of synthesis within which we can make empirical sense of what otherwise appear to be overwhelming quantities of disconnected information. As a bonus, this theory of human social processes also serves as an integrated framework for socio-environmental studies.

Confronting problems in sociology and re-imagining the sociological sciences, it turns out, is a necessary step toward achieving a more accurate view of the socio-environmental territory we have always lived in, but are just beginning to make better sense of. As with any exploration of vast and unfamiliar terrain, a map can be helpful. The next chapter is devoted to developing a map of our socio-environmental context writ large, locating our diverse inquiries in it, and using it to understand the general pattern of socio-environmental processes always at work.

Notes

1. http://www.asanet.org/about-asa/asa-story/what-sociology. Accessed January 21, 2020.
2. https://www.britsoc.co.uk/what-is-sociology/what-do-sociologists-do/. Accessed January 21, 2020.
3. In a recent experiment, these authors submitted 20 hoax papers—"all outlandish or intentionally broken in significant ways"—to respectable peer-reviewed social science and humanities journals (Pluckrose et al. 2018). By the time the story broke, seven had been accepted for publication (with four already published), four had been invited to revise and resubmit, and one was under review. They were taken seriously, others say, "because they fit with social science sub fields in which reason has been exchanged for ideology" (Arden 2018). The architects of the experiment list sociology as one of those fields.
4. From the subject listing in Nature.com, https://www.nature.com/subjects/biological-sciences.

References

American Sociological Association (ASA). 1978. "Is Sociology Relevant to the 'Real' World? Yes, but..." *Footnotes* January: 6–7.

American Sociological Association (ASA). 1989. "Washington University's Sociology Department: An Update." *Footnotes* September: 8–9.

American Psychological Association (APA). 2011. *Annual Report of the American Psychological Association*. http://www.apa.org/pubs/info/reports/2011-report.pdf.

Arden, Rosalind. 2018. "Barking in Academia." https://quillette.com/2018/10/01/the-grievance-studies-scandal-five-academics-respond/. Retrieved February 16, 2020.

Bauman, Zygmunt. 2002. *Society Under Siege*. Cambridge: Polity Press.

Bauman, Zygmunt. 2014. *What Use Is Sociology*, Conversations with Michael Hviid Jacobsen and Keith Tester. Cambridge: Polity Press.

Becker, Howard S., and William C. Rau. 1992. "Sociology in the 1990s." *Society* 30(1):70–74.

Berger, Peter L. 1992. "Sociology: A Disinvitation?" *Society* 30(1):12–18.

Black, Donald. 2000. "The Purification of Sociology." *Contemporary Sociology* 29(5):704–709.

Bolsonaro, Jair. 2019. Bolsonaro's Tweet, in translation. April 26.

Boulding, Kenneth. 1956. "General Systems Theory: The Skeleton of Science." *Management Science* 2(3):197–208.

Burawoy, Michael. 2007. "For Public Sociology." Pp. 23–64 in *Public Sociology: Fifteen Eminent Sociologists Debate Politics and the Profession in the Twenty-First Century*, edited by Dan Clawson, Robert Zussman, Joya Misra, Naomi Gerstel, Randall Stokes, Douglas L. Anderton, and Michael Burawoy. Berkeley, CA: University of California Press.

Burawoy, Michael. Public Sociology website. Accessed September 2016, http://burawoy.berkeley.edu/PS.Webpage/ps.mainpage.htm.

Buttel, Frederick. 2002. "Has Environmental Sociology Arrived?" *Organization and Environment* 15(1):42–54.

Calhoun, Craig, Joseph Gerteis, James Moody, Steven Pfaff, and Indermohan Virk, eds. 2007. *Contemporary Sociological Theory*, 2nd ed. Malden, MA: Blackwell.

Carolan, Michael. 2005. "Society, Biology, and Ecology: Bringing Nature Back into Sociology's Disciplinary Narrative Through Critical Realism." *Organization & Environment* 18(4):393–421.

Carroll, William K. 2013. "Discipline, Field, Nexus: Re-visioning Sociology." Plenary Address at Canadian Sociological Association.

Catton, William R. Jr. 1994. "Foundations of Human Ecology." *Sociological Perspectives* 37(1):75–95.

Catton, William R. Jr., and Riley E. Dunlap. 1978. "Environmental Sociology: A New Paradigm." *The American Sociologist* 13:41–49.

Cole, Stephen. 1994. "Why Sociology Doesn't Make Progress Like the Natural Sciences." *Sociological Forum* 9:133–154.

Cole, Stephen. 2001. *What's Wrong with Sociology?* New Brunswick, NJ: Transaction.

Comte, Auguste. 1848 *A General View of Positivism*. Paris.

Comte, Auguste. 1855. *The Positive Philosophy of Auguste Comte*, translated by Harriet Martineau. New York: Calvin Blanchard.

Conley, Dalton. 2019. *You May Ask Yourself: An Introduction to Thinking Like a Sociologist*, 6th ed. New York: W. W. Norton.

Crossley, Nick. 2011. *Towards Relational Sociology*. Routledge.

Das, Mitra. 1989. "Making Sociology Attractive." *Teaching Sociology* 17:356–360.

Davis, James. 1994. What's Wrong With Sociology?" *Sociological Forum* 9:179–198.

DeCesare, Michael. 2005. "95 Years of Teaching High School Sociology." Paper presented at the Annual Meeting of the American Sociological Association, Philadelphia. http://www.allacademic.com/meta/p18602_index.html.

DeCesare, Michael. 2007. "AP or Not AP? That Is the Question." *Footnotes* December.

Deflem, Mathieu. 2013. "The Structural Transformation of Sociology." *Society* 50(2):156–166.

Donati, Pierpaolo. 2011. *Relational Sociology: A New Paradigm for the Social Sciences.* Routledge.

Elias, Norbert. 1956. "Problems of Involvement and Attachment." *The British Journal of Sociology* 7(3):226–252.
Elias, Norbert. 1978. *What Is Sociology?* translated by S. Mennell and G. Morrissey with a Foreword by R. Bendix. New York: Columbia University Press.
Elias, Norbert. 1987 [1983]. *Involvement and Detachment*, translated by Edmund Jephcott. Oxford: Basil Blackwell.
Elias, Norbert 1991a. *The Society of Individuals*. Cambridge: Basil Blackwell.
Elias, Norbert. 1991b [1989]. *The Symbol Theory*, edited with an Introduction by R. Kilminster. London: Sage.
Elliott, Anthony. 1999. *The Blackwell Reader in Contemporary Social Theory*. Malden, MA: Blackwell.
Elliott, Anthony, and Ray, Larry J. 2003. *Key Contemporary Social Theorists*. Malden, MA: Blackwell Publishers.
Ellwood, Charles. 1907. "Sociology: Its Problems and Its Relations." *American Journal of Sociology* 13(3): 300–348.
Emirbayer, Mustafa. 1997. "Manifesto for a Relational Sociology." *American Journal of Sociology* 103:281–317.
Fabianic, David. 1991. Declining Enrollments of Sociology Majors: Department Responses. *The American Sociologist* 22:25–36.
Fischer, Joern, Robert Dyball, Ioan Fazey, Catherine Gross, Stephen Dovers, Paul R. Ehrlich, Robert J. Brulle, Carleton Christensen, and Richard J. Borden. 2012. "Human Behavior and Sustainability." *Frontiers in Ecology & the Environment* 10(3):153–160.
Flyvbjerg, Bent. 2001. *Making Social Science Matter*. Cambridge: Cambridge University Press.
Foster, John Bellamy. 2002. "Environmental Sociology and the Environmental Revolution: A 25th Anniversary Assessment." *Organization and Environment* 15(1):55–58.
Fox Piven, Frances. 2007. "From Public Sociology to Politicized Sociologist." Pp. 158–168 in *Public Sociology: Fifteen Eminent Sociologists Debate Politics and the Profession in the Twenty-First Century*, edited by Dan Clawson, Robert Zussman, Joya Misra, Naomi Gerstel, Randall Stokes, Douglas L. Anderton, and Michael Burawoy. Berkeley, CA: University of California Press.
Gabriel, Norman, and Stephen Mennell. 2011. "Handing Over the Torch: Intergenerational Process in Figurational Sociology." Pp. 5–23 in *Norbert Elias and Figurational Research: Processual Thinking in Sociology*, edited by Norman Gabriel and Stephen Mennell. Malden, MA: Blackwell.
Geary, Daniel. 2009. *Radical Ambition: C. Wright Mills, the Left, and American Social Thought*. Berkeley: University of California Press.
Ginsberg, Morris. 1937. "Sociology and Human Affairs." Pp. 166–180 in *Human Affairs: An Exposition of What Science Can Do for Man*, edited by

R. B. Cattell, J. Cohen, and R. M. W. Travers, London: Macmillan and Co, Ltd.
Goldthorpe, John. 2000. *On Sociology: Numbers, Narratives, and the Integration of Research and Theory.* Oxford: Oxford University Press.
Grove, Jack. 2015. "Social Sciences and Humanities Faculties 'to Close' in Japan After Ministerial Intervention." *Times Higher Education*, September 14. https://www.timeshighereducation.com/news/social-sciences-and-humanities-faculties-close-japan-after-ministerial-intervention.
Hawking, Stephen, and Leonard Mlodinow. 2005. *A Briefer History of Time.* New York, NY: Bantam Books.
Horowitz, Irving L. 1993. *The Decomposition of Sociology.* New York and London: Oxford University Press.
Huber, Joan. 1995. "Institutional Perspectives on Sociology." *The American Journal of Sociology* 101(1):194–216.
Kalberg, Stephen. 2004. "A Cross-National Consensus on a Unified Sociological Theory? Some Inter-cultural Obstacles." *Perspectives* 27(4):1–9.
Kasper, Debbie. 2011. "Finding Coherence in Sociology: (Finally!) A Foundational Theory." Pp. 121–144 in *New Directions in Sociology: Essays on Theory and Methodology in the 21st Century*, edited by I. Zake and M. DeCesare. Jefferson, NC: McFarland.
Kasper, Debbie. 2014. "Codifying Figurational Theory and Mapping Common Ground in Sociology…and Beyond." *Human Figurations* 3(1). http://hdl.handle.net/2027/spo.11217607.0003.104.
Kasper, Debbie. 2016. "Re-conceptualizing (Environmental) Sociology." *Environmental Sociology* 2(4):322–332.
Keith, Bruce. 2000. "Taking Stock of the Discipline: Some Reflections on the State of American Sociology." *The American Sociologist* 31:5–14.
Keith, Bruce, and Morten G. Ender. 2004. "The Sociological Core: Conceptual Patterns and Idiosyncrasies in the Structure and Content of Introductory Sociology Textbooks, 1940–2000." *Teaching Sociology* 32:19–36.
Kolbert, Elizabeth. 2006. *Field Notes from a Catastrophe: Man, Nature, and Climate Change.* New York: Bloomsbury.
Lane, Charles. 2012. "Congress Should Cut Funding for Political Science Research." *Washington Post*, June 4. https://www.washingtonpost.com/opinions/congress-should-cut-funding-for-political-science-research/2012/06/04/gJQAuAJMEV_story.html.
Lieberson, Stanley, and Freda B. Lynn. 2002. "Barking Up the Wrong Branch: Scientific Alternatives to the Current Model of Sociological Science." *Annual Review of Sociology* 28:1–19.
Mann, Michael. 1986. *The Sources of Social Power*, Vol. 1. Cambridge: Cambridge University Press.

Massey, Douglas. 2007. "The Strength of Weak Politics." Pp. 145–157 in *Public Sociology: Fifteen Eminent Sociologists Debate Politics and the Profession in the Twenty-First Century*, edited by Dan Clawson, Robert Zussman, Joya Misra, Naomi Gerstel, Randall Stokes, Douglas L. Anderton, and Michael Burawoy. Berkeley, CA: University of California Press.

Matthews, Dylan. 2014. "Why Congressional Republicans Want to Cut Social Science Research Funding." *Vox*, November 12. http://www.vox.com/2014/11/12/7201487/congress-social-science-nsf-funding.

Michalski, Joseph. 2008. "Scientific Discovery in Deep Social Space: Sociology Without Borders." *Canadian Journal of Sociology* 33(3): 521–553.

Mole, Beth. 2013. "NSF Cancels Political-Science Grant Cycle." *Nature News*, August 2. http://www.nature.com/news/nsf-cancels-political-science-grant-cycle-1.13501.

National Academies. 2006. "Taxonomy of Fields and Their Subfields." http://sites.nationalacademies.org/PGA/Resdoc/PGA_044522.

National Institutes of Health, OBSSR. n.d. "BSSR Definition." https://obssr.od.nih.gov/about-us/bssr-definition/.

OECD. 2004. "Re-inventing the Social Sciences." Organisation for Economic Co-operation and Development. http://www.oecd.org/science/sci-tech/33695704.pdf.

Pearman, William A. 1992. "The Invisibility of Sociology: The State of Our Discipline." *Footnotes* 20(2):8.

Phillips, Bernard. 1999. "Confronting Our Tower of Babel." *Perspectives* 21(4):2–6.

Pluckrose, Helen, James A. Lindsay, and Peter Boghossian. 2018. "Academic Grievance Studies and the Corruption of Scholarship." *Areo*, October 2. https://areomagazine.com/2018/10/02/academic-grievance-studies-and-the-corruption-of-scholarship/. Retrieved February 16, 2020.

Quilley, Stephen. 2010. "Integrative Levels and 'the Great Evolution': Organicist Biology and the Sociology of Norbert Elias." *Journal of Classical Sociology* 10(4):391–419.

Quilley, Stephen, and Steven Loyal. 2005. "Eliasian Sociology as a 'Central Theory' for the Human Sciences." *Current Sociology* 53(5):809–830.

Rabesandratana, Tania. 2014. "E.U. Commission Wants to Divert Horizon 2020 Money into New Investment Fund." *AAAS*, November 27. http://www.sciencemag.org/news/2014/11/eu-commission-wants-divert-horizon-2020-money-new-investment-fund.

Raskin, Paul. 2016. *Journey to Earthland: The Great Transition to Planetary Civilization*. Boston: Tellus Institute.

Riley, William. 2016. "Behavioral and Social Sciences: Multiple Disciplines That Explain How Humans Influence Health and Wellbeing."

February 23. https://obssr.od.nih.gov/behavioral-and-social-sciences-multiple-disciplines-that-explain-how-humans-influence-health-and-wellbeing/.
Ritzer, George, and Douglas J. Goodman. 2008. *Sociological Theory*, 7th ed. Boston: McGraw-Hill.
Rule, James B. 1994. "Dilemmas of Theoretical Progress." *Sociological Forum* 9:241–257.
Salter, Frank. 1996. "Sociology as Alchemy." *Skeptic* 4(1):50.
Salter, Frank. 2012. "The War Against Human Nature in the Social Sciences." *Quadrant Magazine* 56(6):49–57.
Seidman, Steven. 1991. "The End of Sociological Theory: The Postmodern Hope." *Sociological Theory* 9(2): 131–146.
Small, Albion. 1905. *General Sociology.* Chicago: University of Chicago Press.
Smith-Lovin, Lynn. 2007. "Do We Need a Public Sociology?" Pp. 124–134 in *Public Sociology: Fifteen Eminent Sociologists Debate Politics and the Profession in the Twenty-First Century*, edited by Dan Clawson, Robert Zussman, Joya Misra, Naomi Gerstel, Randall Stokes, Douglas L. Anderton, and Michael Burawoy. Berkeley, CA: University of California Press.
Stinchcome, Arthur. 1994. "Disintegrated Disciplines and the Future of Sociology." *Sociological Forum* 9:279–291.
Stinchcombe, Arthur. 2007. "Speaking Truth to the Public, and Indirectly to Power." Pp. 135–144 in *Public Sociology: Fifteen Eminent Sociologists Debate Politics and the Profession in the Twenty-First Century*, edited by Dan Clawson, Robert Zussman, Joya Misra, Naomi Gerstel, Randall Stokes, Douglas L. Anderton, and Michael Burawoy. Berkeley, CA: University of California Press.
Turner, Jonathan H. 2005. "Is Public Sociology Such a Good Idea?" *The American Sociologist* 36:27–45.
Turner, Jonathan H. 2006. "American Sociology in Chaos: Differentiation without Integration." *The American Sociologist* 37(2):15–29.
Turner, Jonathan H. 2008. "The Practice of Scientific Theorizing in Sociology and the Use of Scientific Theory in Sociological Practice." NCSA Plenary Address, published in *Sociological Focus* 41(4):281–299.
Turner, Stephen P., and Jonathan H. Turner. 1990. *The Impossible Science: An Institutional Analysis of American Sociology*. Newbury Park, CA: Sage.
Van Den Bergh and Godfried Van Benthem. 1986. "The Improvement of Human Means of Orientation: Towards Synthesis in the Social Sciences." Pp. 109–135 in *Development Studies: Critique and Renewal*, edited by Raymond Apthorpe and András Kráhl. Leiden: E. J. Brill.
Wallerstein, Immanuel. 1991. *Unthinking Social Science.* Cambridge: Polity Press.
Wallerstein, Immanuel. 1996. *Open the Social Sciences: Report of the Gulbenkian Commission on the Restructuring of the Social Sciences.* Stanford: Stanford University Press.

CHAPTER 4

Mapping the Territory

"Where would we be without maps?" asks Jerry Brotton, author of *A History of the World in 12 Maps*. "The obvious answer," he quips, is "lost" (2012:4). According to a generally accepted definition, maps are "graphic representations that facilitate a spatial understanding of things, concepts, conditions, processes, or events in the human world" (Harley and Woodward 1987:xvi). Replacing the word spatial with relational helps us see how theories, and the visual models that depict them, can function like maps.

Without the benefit of a map in the form of a reliable general theory of human social life there's been a great deal of wandering and wondering. Many have come to doubt the very possibility of a theoretical framework that can contain and accurately depict the general patterns of relations among the key things, concepts, conditions, processes, and events in the human realm. Nevertheless, there have been others who insist on the possibility, though not without acknowledging its challenges. "It must be admitted," one British sociologist in the 1930s said, "that the body of facts accumulated by the specialists is so vast that the systematic sociologist can only approach it with fear and trembling. Sociology still awaits its Newton or its Darwin" (Ginsberg 1937:180). More than 50 years later, sociology was still awaiting "the great systematizers…who will establish a common language and overarching frame for the analysis of social life" (Zald 1991:173).

D. Kasper, *Beyond the Knowledge Crisis*,
Palgrave Studies on Norbert Elias,
https://doi.org/10.1007/978-3-030-48370-8_4

The difficulties associated with the prospect of grand social theory, this chapter argues, are largely due to confusion about humans' place in the world, which is reinforced by the faulty concepts and models used to think and speak about it. The first step in clearing up that confusion is to locate people, their activities, and all social phenomena firmly in their native ground, the physical and biological processes they happen in and through.

We begin by reviewing some of the most influential ideas in the west about how and where people fit in the grand scheme of things, with a special focus on scholarly attempts to unite humans and nature. Despite good intentions, a false separation persists. As with biology, the advancement of which depended on a much vaster sense of time, achieving a more accurate understanding of human social life requires a more expansive view. This chapter guides readers through a process of conceptually zooming out to be able to imagine a bigger picture. With sufficient distance, the mental equivalent of a satellite, we can discern some overall patterns and processes that we cannot see from the ground. From here, we can create a map of the socio-environmental territory we have only just begun to explore. It will be at a small-scale,[1] what, in theoretical terms, Elias described as a high level of synthesis. Distinct in its explicit encompassing of the three basic levels of phenomena, this sociological map serves, at the same time, as an integrative theoretical framework for socio-environmental studies.

Putting People on the Map

World maps have always been more than just devices to note landmarks and help people find their way around. Throughout history, efforts to map the known world have represented not only the physical characteristics of a territory, but complex interactions between those and available science and technology, aspects of the cartographer's worldview and travel experiences, and the economic, political, and religious priorities of the time. In short, Brotton shows, worldviews exert important influences on shaping world maps, which in turn define, fortify, and alter worldviews.

Pictures meant to help us visualize how humans relate to non-human phenomena operate in much the same way. Like maps, their purpose is orientation, but they too reflect and perpetuate certain ideas about reality held by people at a given place and time. Featured here are some of the "maps" that have had an especially powerful influence on our sense of where people fit in the grand scheme of things.

Above and Outside of Nature

In Western thought, the notion of "the great chain of being" has had a lasting influence on our ideas of humanity as distinct from and superior to nature. Figure 4.1 depicts the hierarchical structure of beings and matter believed in medieval Christianity to be decreed by God. In rank order from the top down are: God, angels, royalty, nobility, commoners, domesticated and wild animals, plants, and non-living things.

Fig. 4.1 The Great Chain of Being (From *Retorica Christiana*, published by Diego Valdes in 1579. Public domain, Wikipedia)

Long after this idea's legitimacy waned in the age of science, the notion that humans are dominant over or separate from the rest of nature persists in some forms of religious thought and secular domains, including academia. Note the neglect of biophysical realities, for example, in conventional economics. The "circular flow" model of goods and services in market economies (see Fig. 4.2) depicts a continual cycling of inputs and outputs, appearing to come from and go nowhere, with no entropy, waste, or negative consequences. This is analogous, ecological economist Herman Daly argues, to biologists depicting animals with circulatory systems but no digestive tracts. Just as an animal's digestive system connects it with its environment at both ends, human economic activities depend on inputs of stuff *from somewhere* and generate outputs that *go somewhere*. Despite its obvious shortcomings, the idea remains a staple in economics education, perpetuating a distorted sense of the consequences of economic activities. There are parallels in other humanities and social science disciplines, which have "focused on 'the human' in a way that has often excluded or backgrounded the non-human world," as stated in the inaugural issue of *Environmental Humanities* (Rose et al. 2012:1).

As a response to growing awareness of this exclusion, the emergence of "environmental" or "ecological" subdisciplines in the humanities

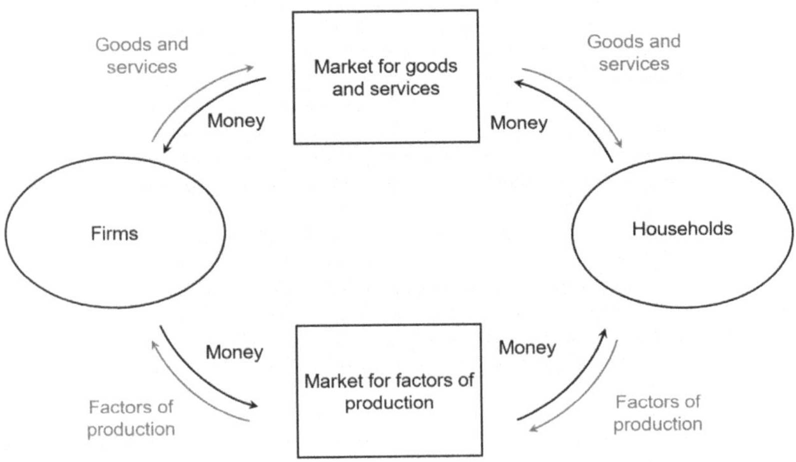

Fig. 4.2 Economics' basic circular flow model

and social sciences in the late twentieth century represented significant progress. For the most part, though, they remain somewhat marginal to their respective mainstreams and did not effect the deeper disciplinary transformations that were, in some cases, intended. Despite ongoing efforts along these lines, the conclusion consistently reached is that we must do better, especially in terms of more effectively integrating the "social" and "natural" sciences (Moran 2010; Smith 2009; Stafford et al. 2010; Stock and Burton 2011; Victor 2015; Zax 2009).

Linking Humans and Nature

Beyond the confines of what environmental subspecialties tried to accomplish *within* their own disciplines, there was a wave of efforts to more broadly depict the overall nature of relationships among all biophysical and human social processes. The ultimate goal was a "global view of the earth system that emphasizes the connectedness of all intervening parts" (Roederer 1988:664). This was no easy task. In the wake of steadily intensifying specialization, the sciences which studied the Earth and ecological systems worked mostly independently from one another and were even further removed from the social sciences. It would require significant and highly coordinated efforts to move things in the other direction.

Undaunted, some took up the task. One such endeavor was the International Geosphere-Biosphere Program (IGBP). In 1983, the National Academy of Sciences sponsored a workshop to consider the development of a new program on the "study of whole systems of interdisciplinary science in an effort to understand global changes in the terrestrial environment and its living systems" (Commission 1983). The integration of different kinds of sciences was simultaneously the novel promising element and a source of tension in this uneasy new partnership. For one thing, those in the life sciences noted that, although the project's main goal was to preserve the global life-supporting system, the process was being dominated by the physical sciences (Solbrig 1985). Another point of contention was the outright exclusion of the social sciences during the program's formulation in 1984 (Mooney et al. 2013:3665). After further deliberation, in 1986 the planning group announced its intent to launch IGBP, stating explicitly that the domain of social science would not be included (Greenway 1996).

During the same decade, a parallel program was being developed by NASA which culminated in some landmark reports on the functioning of

Earth systems and an accompanying visual diagram. These projects, led by meteorologist Francis Bretherton and published in 1988, portrayed global change at two different timescales: thousands to millions of years, and decades to centuries. Unlike the IGBP, the Bretherton diagrams *did* include human activities in their map of physical and biospheric processes. Their representation, however, was grossly disproportionate, with "human activities" in both versions relegated to a single oval off to one side (see Figs. 4.3 and 4.4). In the end, the still-developing IGBP project ended up adopting a streamlined version of the short-term Bretherton model, omitting the human drivers altogether. In terms of putting people on the map, so far, not so good.

Despite the poor start, there were positive developments. One was greater clarity about the attitudinal, structural, methodological, and conceptual obstacles to forging the badly needed connections between the social and natural sciences (Kates 1985; Miller 1989). Another was

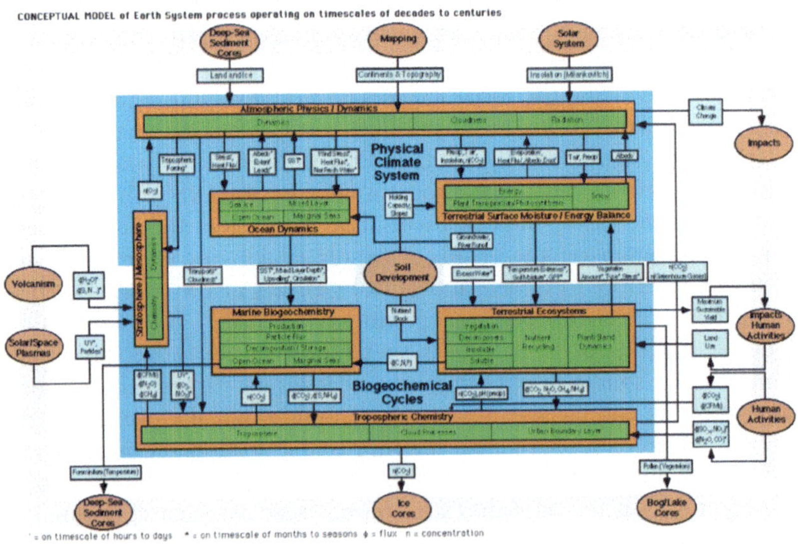

Fig. 4.3 The Bretherton Diagram. A conceptual model of the functioning of the Earth system in time scales of thousands to millions of years (From *Earth System Science, Overview: A Program for Global Change*, 1986. Image used with permission of National Academies Press)

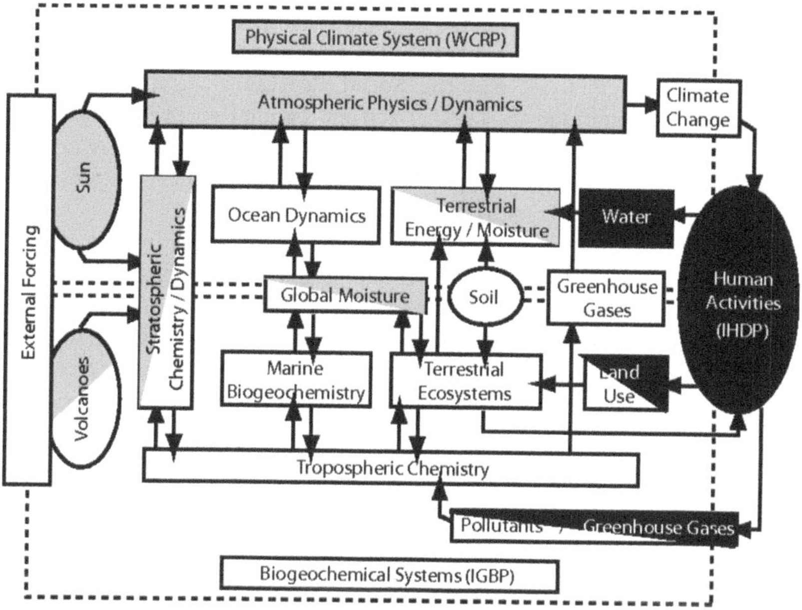

Fig. 4.4 The simplified Bretherton Diagram. A conceptual model of the functioning of the Earth system in time scales of decades to centuries (From *Earth System Science, Overview: A Program for Global Change*, 1986. Image used with permission of National Academies Press)

the emergence of new efforts to incorporate social science into the study of global change.

In 1986, the International Social Science Council (ISSC) and its Human Dimensions Program initiative resolved to create a committee to explore the possibility of developing an international social science research program that would be analogous to and complement the emerging IGBP. Formally proposed in 1990, the goal was to explicitly study the human dimensions of global environmental change, including: resource use, environmental perception, impacts of economic and political structures at multiple levels, land use, energy conversion and consumption, industrial growth, and sustainable development. A few years later, this new effort partnered with the Consortium for International Earth Science Information Network (CIESIN, established in 1989 for the purpose of providing information to help scientists, decision-makers, and the public better understand the relationship between humans and the environment) for help maintaining data and information systems. One result of their work was the Social Process Diagram, a conceptual framework intended as a social science parallel to existing physical science-based models of global change. It featured interactions among population, economic systems, production systems, political systems, and human behavior (see Fig. 4.5).

What the Bretherton diagram did with human activities, the Social Process Diagram did with environmental processes, relegating them mostly to a single box at the bottom. So, although it usefully highlighted significant social dimensions of global environmental change, it didn't add much to our understanding of the relationships between them. This effort to create human dimensions parallel to the IGBP program initially failed, in part because of international organizational dynamics and funding issues. In 1996, however, the Human Dimensions Program got a second life, as the International Human Dimensions Program on

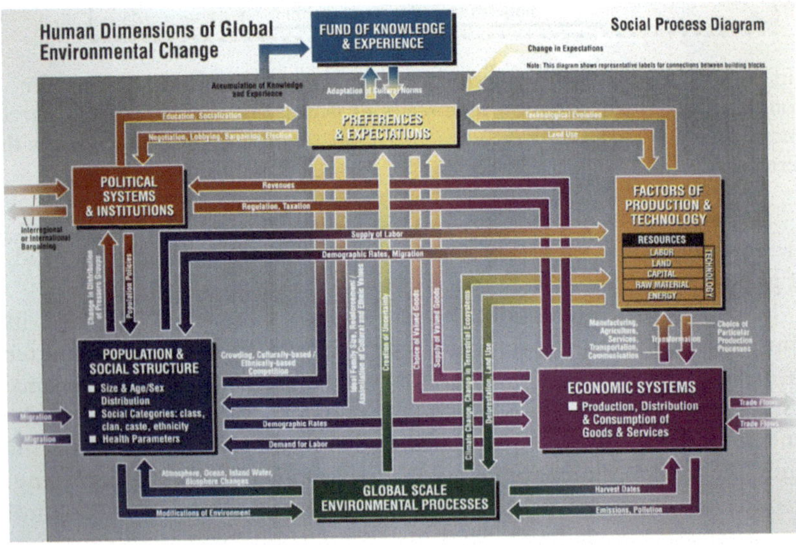

Fig. 4.5 The Social Process Diagram (From *Pathways of Understanding: The Interactions of Humanity and Global Environmental Change*, 1992. Image used with permission of CIESIN)

Global Environmental Change. The original seven priorities were reduced to three: institutional dimensions of global environmental change, industrial transformation and global environmental change, and human security and global environmental change. The program was jointly sponsored by the International Social Science Council and International Council on Science, setting the stage for more meaningful partnerships between social and natural scientists in building integrated systems models (Ehlers and Kosinski 1998; Mooney et al. 2013).

In the end, social science was increasingly brought into natural science programs, as evident in the formation of the Intergovernmental Panel on Climate Change assessment process, the Millennium Ecosystem Assessment, Sustainability Science, the Resilience Alliance, and Future Earth (Mooney et al. 2013). The highlighting of environmentally significant social factors did get some people thinking about the biophysical impacts of social systems in ways they hadn't before, but the practical value of these models for teaching, research, and action proved quite limited.

One reason is their visual complexity. The overwhelming number of boxes, terms, and crisscrossing arrows makes it difficult to interpret and explain key relationships and processes. While there's a place for very detailed models, the details can best be made sense of in the context of a more universal framework, which did not exist. This relates to another problem, their limited historical scope. Ingredients like regulation, taxation, lobbying, and manufacturing can be helpful for drawing attention to the significant factors currently at play in some settings, but the present-centeredness of models rendered them unsuitable for understanding socio-environmental processes more generally.

Since that time, models attempting to depict the relationship between human activities and environmental processes from a higher level of synthesis have been increasingly introduced. Ironically, though, their efforts to conceptually link these different phenomena often reinforce a sense of separateness. The most common structure, consisting of two clear sides with humans or societies on one and nature or environment on the other, only reinforces the division. Despite the helpful arrows drawn between them to indicate dynamism, connections, and "coupling," the nature of those linkages remains mysterious, as do authors' choices about what components get included in these frameworks, which are most often presented without justification (Pulver et al. 2018).

These and similar efforts represent progress in the overall attempt to put people on the map of biophysical phenomena. Still, Robert Chen

(2013) of CIESIN points out, "the need to better characterize and understand the complexity of interactions between social structures and processes and the environment remains a significant challenge." Attempts to characterize overall socio-environmental systems have not yielded a model that is at the same time clear, comprehensive, integrative, general, and actionable. In defense of those who have tried, this is a pretty tall order. Making matters worse are habits of thought and speech that constrain our ability to accurately think and communicate about the dynamic interrelatedness of socio-environmental phenomena.

Perhaps over time, Chen allows, "we will be in a better position to create an improved social process road map" to understand how we got here and to guide the development of a sustainable future. There's not much chance of that, however, if we continue to work primarily from specialized points of view. In socio-environmental studies, some say, there still seem to be two separate audiences, "natural scientists…who want to include in their world model a box called 'people'" and "social scientists interested in resource issues who wish to include in their model a box called 'natural environment'" (Westley et al. 2002:103).

So what *are* we to do? Dealing effectively with the complex problems facing us absolutely demands a better appreciation of the nature of socio-environmental processes. That requires interdisciplinarity. The trouble is, physicist Juan Roederer pointed out, "we are ill prepared for it. Our students are ill prepared for it. Our government organizations are ill prepared for it. The international scientific establishment is ill prepared for it" (1988:666).

The recognition that knowledge structures have not kept pace with the challenges we face inspired some efforts toward a re-organization of those structures. But the severe mismatch between "the slowness of the process in relation to the need for restructuring," political scientist Lynton Caldwell contends, has delayed real action (1983:256). We cannot afford further delay. With a note of optimism, Caldwell went on to suggest, the necessary restructuring process *could* be accelerated if a feasible solution can be visualized. This brings us back to the prospect of a map to help us envision our socio-environmental territory, the creation of which requires a much vaster view.

Satellite View

Even after mapmakers had been able to use triangulation and surveying techniques to calculate distances and create fairly accurate renderings of the world's coastlines, continents, and countries, humankind sought glimpses of things from above. Key moments in my own nation's history exemplify that striving for ever-higher birds-eye views. In 1935, the United States launched the Explorer II, a high-altitude balloon ascending 13.7 miles, a height from which the Earth's curvature can be detected (see Fig. 4.6). At the time, these were the highest pictures ever taken of Earth, until 1947 when scientists sent guided ballistic missiles armed with cameras more than 100 miles above the Earth's surface. Pasting frames together, they created a panorama covering more than a million square miles at a glance and clearly showing the contrast between Earth and the blackness of space it was set against (see Fig. 4.7).

Fig. 4.6 Explorer II, 1935 (*Credit* Richard Hewitt Stewart/National Geographic Creative)

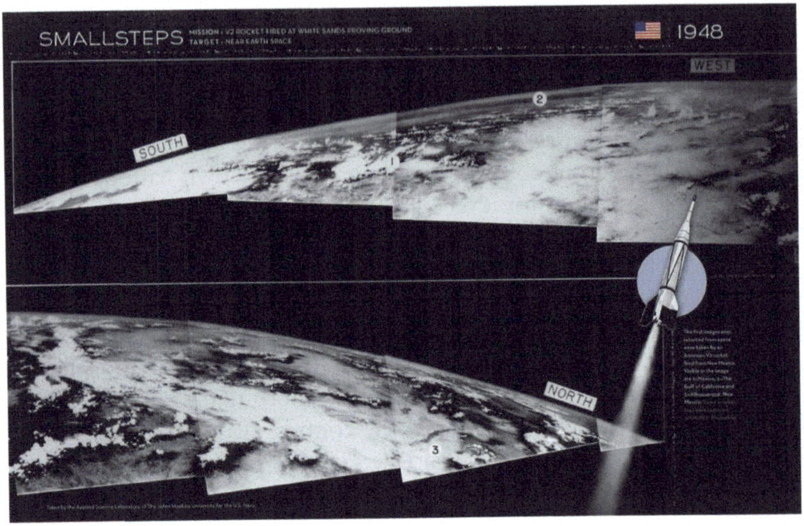

Fig. 4.7 First Pictures of Earth from 100 Miles in Space, 1947 (Image used with permission of NASA)

This view was enhanced a couple decades later when, in 1968, the crew of Apollo 8 caught a glimpse of the Earth "rising" from a quarter million miles away while their craft orbited the moon (see Fig. 4.8). Capturing in vivid color the finite, unique, and isolated qualities of the planet, this striking image graced the cover of Life's *100 Photographs That Changed the World*, in which photographer Galen Rowell described it as the most influential environmental photograph ever taken. And in 1972, just five hours and six minutes after the launch of the Apollo 17 mission and from a distance of 28,000 miles (about 45,000 km), the crew snapped one of the most widely shared photographic images in existence. In it, the Earth is almost fully illuminated, with several features clearly visible, including the polar ice cap, the Arabian Peninsula, the coastline of Africa, and Asia on the horizon (see Fig. 4.9).

Since the publication of these initially groundbreaking photos, images of the Earth from space have become so commonplace that it's difficult to imagine a time when people didn't see our planet like this. Whether we know it or not, the ability to visualize Earth in its larger context has changed our thinking in important ways. Those who have

Fig. 4.8 *The Blue Marble*, 1968 (*Image credit* NASA)

gazed upon it directly invariably describe the experience as transformative. Using adjectives like small, beautiful, isolated, fragile, vulnerable, peaceful, whole, they describe qualities of our world that are impossible to see or fully appreciate from the ground. Perhaps more importantly, James Lovelock said of these images, they have "given rise to a whole new set of questions and answers" (1979:8). And they continue to. The satellite imagery NASA makes freely available, for example, allows us to observe cloud cover, fire, land surface temperature, snow cover, and other features of global systems and how they change over time, prompting new hypotheses, investigations, and findings.

Fig. 4.9 *Earthrise*, 1972 (*Image credit* NASA)

Similarly, a satellite-like perspective on socio-environmental processes would enable researchers to more easily situate their work in relation to others', clarifying connections and raising new questions. But because no camera can capture an image of *these* processes, we have to create our own picture. We begin by gaining enough mental distance to be able to detect the overall patterns (at a high level of synthesis) at play. Working with the outline they provide, we can use existing knowledge to connect more dots and fill in the details of our map. For a map to be useful, though, it must be founded on accurate assumptions about basic starting conditions. In this case, that means it must account for and be consistent with the fundamental laws operating at physical, biological, and social levels.

Laying Down the Laws

The notion of physical "laws" has been with us for a few centuries; by now, they are well established. Only more recently have formal laws begun to take hold in biology. Wikipedia lists a couple dozen; others propose just three (Trevors and Saier 2010). These are not quite as familiar as some of the more famous laws of physics, but those who know them take them to be quite sound. As we know, early attempts to establish sociological laws didn't fare so well.

One problem was that an eagerness to prove the legitimacy of the social realm as a subject in its own right led to a portrayal of social phenomena as autonomous and the resulting widespread neglect of its biophysical contexts. In fairness, at that time some of the knowledge (about environmental conditions, human evolution, and the brain, for instance) upon which a more accurate understanding of social phenomena depended was in its infancy. Regardless of the reason, without sufficient grounding in relevant scientific knowledge sociology was unable to develop reliable universal "laws" of its own. Eventually, most came to reject the very possibility, citing the overwhelming complexity and diversity conceivable in the spectrum of human experience as one of the chief reasons.

Not so fast, others say. Biological evolution involves an extravagantly diverse and complex arrays of causal conditions, yet biology has managed to progress. It has done so precisely *because* the study of evolution is structured according to "a modest number of *principles* rather than a virtually infinite number of *possible variables*" (Lieberson and Lynn 2002:8). It is far more promising, they argue, to be able to identify a variety of mechanisms, stated in generalized form as principles or theories, which can be called upon to explain any instance of evolution among life, anywhere on Earth. The explanatory power of these biological principles is widely accepted.

As yet, there is no such acceptance that sociology could ever explain the spectacular diversity of social life across the world and history. In this context, the quest for "the sociological theory applicable to every conceivable instance of anything across the social universe, past, present, and future" (Black 2000:705), appears ludicrous to many. There is, however, a small minority who continue to insist that not only is it not ludicrous, it's within our reach. The question for them is not whether we need a general theory (we do), but how to get it. The answer, I argue here, lies in explicitly aligning sociological knowledge with knowledge about phenomena at other levels, especially biology, "the next level down."

Although human happenings are not reducible to biological phenomena, an appreciation for how key biological traits influence human social life is indispensable. Without that, our subject matter is left hanging in the air, necessitating the invention of "skyhook" concepts which, as we saw in Chapter 3, have caused serious problems. The complexity of modern sociological theories, Elias assures us, is due "not to the complexity of the field of investigation which they seek to elucidate, but to the kind of concepts employed" (1978:111). In particular, more appropriate core concepts must reflect and be commensurate with knowledge at other levels. Conveniently, using that knowledge to inform a modest set of scientifically grounded sociological principles both highlights the shortcomings of certain conventional concepts and points us toward more adequate alternatives. Four such premises are discussed below.

1. Humans are biological organisms.

This one may seem too obvious to need mentioning, yet somehow, we often manage to forget that our very existence depends on the physical, chemical, and biological processes constantly at work within and between ourselves and the world. Whatever our reasons for forgetting, the distorted maps of reality we end up with have led us into one dead end dichotomy after another (e.g., individual/society, culture/nature, micro/macro). The simple and easily observable proposition that, whatever else we might be, we are definitely biological organisms, serves as a basis for ground-truthing the nature of our existence and finally displacing the human exemptionalism paradigm that still prevails.

Like other organisms, our lives are characterized by ongoing interactions with the world. Our survival depends on getting the right stuff in the right quantities and combinations at the right times. Our very bodies are made of the world's elements—oxygen, carbon, hydrogen, nitrogen—and are constantly combining and recombining them in the innumerable forms required for the structure, function, and regulation of tissues, organs, systems, and so on. In order to maintain necessary balances, our bodies continually move materials in and out. We breathe, drink, sweat, absorb, eat, and excrete. As long as we're alive, and for some time afterward, we're engaged in a complicated exchange with the world's biophysical processes. As such, our existence and activities have biophysical impacts, a fundamental condition we share with all living beings.

Historically, there's been resistance to explicit recognition of our biological nature, as if it were somehow demeaning or reductive. To the contrary, not only does acknowledging our biological nature *not* diminish us, it gives us even *more* to marvel at. Composed of star stuff, astronomer Carl Sagan reminded viewers of his television series, we have become "a way for the cosmos to know itself." What an extraordinary evolutionary journey!

Along the way, the organizational patterns operating in and among life forms have become differentiated according to genetic instructions and the environmental influences on them at multiple scales, inside and out. One noteworthy divergence along those lines was the development of social behavior. From bees and bats to wolves and whales and so much more in between, these adaptations proved beneficial to survival and reproduction in particular contexts, and were thus passed on to become defining traits of some species. As one of them, *Homo sapiens* are inherently social beings.

2. Humans are social organisms.

When it comes to "being social," many assume it's a choice we have, or a personality trait present in degrees, but that's not what's meant by the phrase here. It represents the fact that humans are social *by nature*, along with many other species. Over the roughly six million years of hominid evolution, apes and their descendants have always lived socially in groups. *Homo sapiens*, however, took social life to a whole new level, for a couple of key reasons.

For starters, more than any other organism, early human life is characterized by an exceptionally long period of dependence. The cranial expansion that accompanied larger brains meant that for a mother's body to be able to accommodate a baby's head in birth, babies must be born at an earlier stage of development. In colloquial terms, we might say that, compared to other animals, humans are born "half-baked" and therefore need protracted intensive care. As a result, mothers too historically relied on a network of others for assistance, especially during the more physically demanding periods of late pregnancy, birth, and infant care. Additionally, beyond the usual things we associate with a baby's survival (e.g., milk, warmth, protection), human infants need social interaction. As (mostly) unintended experiments have illustrated, this is not a luxury, but an essential requirement of our upgraded nervous systems which can only develop normally in a context of physical and psychological contact with others.

One might reasonably wonder why evolution would favor such a costly trait. Remembering that evolution is largely a matter of trade-offs, we can conclude that the price of prolonged dependency was worth the advantages conferred by our larger brains. Over evolutionary time, larger cranial capacity in primates correlates with expansions in social intelligence, a key that unlocked the gate to new horizons of developmental potential. In chimpanzees and Australopithecines, this translated into the ability to manage greater numbers of relationships. Subsequently, in *Homo habilis*, *Homo erectus*, *Homo sapiens*, and other species, it meant increasingly complex emotional and communicative abilities for managing community life, which formed the basis of culture, a revolutionary means of adaptation independent of biological evolution (Massey 2005).

The mimetic culture of *Homo erectus and Homo neanderthalis*, for example, led to a variety of important new social structures, provided a new vehicle for social control and coordination, and served as the cognitive basis of teaching skills and cultural innovation. This powerful adaptive tool enabled these species to branch out from Africa to other locales with a wide range of habitats. Still, in over a million years, they never expanded beyond Southern Europe and Asia, while *Homo sapiens* (henceforth referred to simply as humans) populated nearly the entire planet in just 50,000 years. How was it that our species could survive and multiply so spectacularly in such diverse environments? The answer, it is believed, lies with certain evolutionary changes in our brains.

3. Humans' exceptionally dynamic brains make them exceptionally changeable beings.

A particular development in the evolution of our brain allowing for radical changeability, what neuroscientists call "neural plasticity," made possible a revolutionary kind of being and a revolutionary way of life. This ability for the brain to change itself is not unique to humans, but the extraordinary degree of change possible is.

This plasticity is due to ongoing processes of formation, pruning, and reconfiguration of synaptic structures, connections between neurons, and entire neural networks in response to our environments, circumstances, and experiences. It is thought to be the evolutionary result of changes in a regulatory gene that determines when the process of embryonic neural cell division will stop. The process in humans goes on for enough rounds to give us about 100 billion neurons, whereas in other species it stops sooner (Doidge 2007). The number itself is less important than the connections it makes possible, where the real action of interest to us takes place. More neurons mean exponential increases in possible connections. If we were to consider the number of possible neural circuits in the human brain, scientists have calculated, "we would be dealing with hyper-astronomical numbers: 10 followed by at least a million zeros," which is vastly larger than the number of particles in the known universe (Edelman and Tononi 2000:38)!

We may be able to quantify potential connections, but the mental capacities they afford appear to be incalculable. The important distinction these changes in our brains brought about was the gradual shift toward the prominence of learned over unlearned behaviors. The result in humans is, what Elias calls, "an evolutionary breakthrough" beyond which conduct is directed less by instincts and innate drives and more by impulses shaped by experience and social learning (1987:344). In other words, we come into the world with less genetic information directing our behaviors and instead rely largely on social learning (via imitative and empathetic abilities and the use of symbols and concepts as units of meaning) to acquire the knowledge and skills which enable us to feed and protect ourselves, reproduce and rear young, and do the many other things being a member of a society entails. The crucial point here is that humans not only *can* learn, store, and communicate information with others, we *must* do so.

Babies and children need to learn certain things at certain times to ensure normal physiological and neurological development. The capacity to perceive stereoscopic depth, for instance, requires early experience with binocular vision, which has implications for perceptual and cognitive development. Similarly, variation in one's tonal environment is needed to develop the capacity to perceive a range of tones, on which language processing and proficiency is eventually built (Fox et al. 2010). At certain phases of human development, our openness is greater and the rate of change more intense, but we're learning all the time, whether we realize it or not. Because of that, we're always changing—in ways large and small, visible and invisible. Interestingly, the processes responsible for this constant change are also involved in creating stable and sometimes rigid patterns. At less conscious levels, the neural pathways and networks engaged in our repeated behaviors get reinforced, making us better at doing them and thus more likely to repeat them. Then again, "even when we do the same behavior day after day, the neuronal connections responsible are slightly different each time because of what we have done in the intervening time" (Doidge 2007:208). The ongoing process of configuring and reconfiguring neural connections helps explain both our apparent stability *and* enduring openness to change.

Understanding all this, we can see that, contrary to popular misconceptions, human social learning, commonly called "nurture," is not separate from or in opposition to "nature." It is mandated by, intertwined with, and emerges from it. This insight resolves the supposed paradox of humans' "elaborate genetic innateness and elaborate open-ended flexibility" (Wilson et al. 2014:413). While learning processes themselves are biologically based, the content of what is taught and learned is almost infinitely variable. Consider, for example, our capacity to learn to speak, read, and write a language; hunt, tan and sew hides; extract fiber and weave textiles, drive a horse, carriage, or car; build shelters, do math, resolve conflicts, use medicines to heal and weapons to kill, manage relationships, feel brand loyalty, adapt to ever-changing computer software, navigate the internet, use a cell phone, change a diaper, associate gendered meanings with the sexes, love or hate certain categories of people, prefer a fashion, hold religious beliefs and political ideologies, and on and on. It's the socially shared understanding of all that content—stored in symbols, language, beliefs, values, norms, and material artifacts—which makes up culture. As the emergent property of human biology enabling us to adapt and "evolve" more rapidly than other organisms without having to wait on biological evolution, culture was the real breakthrough.

Knowing this is immensely empowering. Only with an understanding of how our habits of mind and body form can we shape them with greater awareness. Central to that understanding is our interdependence with others. Without a biologically pre-packaged repertoire of information and instincts, it is only through relations with others that we acquire the fundamental operating systems through which we can interpret and engage with the world.

4. Humans exist in mutual relations of interdependence.

As biological organisms, we are inherently interdependent with the biophysical phenomena that comprise us and the contexts of our lives. Thanks to our changeable brains, this holds at the social level too. Beyond our very existence and the preexisting relational structures we enter at birth, is the fact that it is only through relations with others (initially direct and immediate, but potentially expanding to include indirect relations over greater distances of space and time) can one acquire language, knowledge, skills, habits, preferences, ideas, insights, tensions, and all other manner of ingredients with which one formulates an identity. Importantly, this situation in no way precludes individual uniqueness.

People inevitably differ according to their natural constitutions (via genetics and developmental conditions), relational positions, and personal experiences. Without a doubt, this distinctive makeup has an indelible influence throughout the course of one's life. Equally irrefutable, however, is that the way a child's malleable features gradually take firmer shape as he or she grows "*never* depends solely on his constitution and always on the nature of the relations between him and other people" (Elias 1991:22).

On top of all that, humans' extravagant information processing systems further implicate us in relations of interdependence. The gains made in flexibility of learning and behavior were paid for with a loss in stability. Whereas extinguishing the information a species stores and transmits genetically requires the elimination of every individual in that species everywhere, losing knowledge in a human culture requires only one generational break in its transmission, making cultural knowledge somewhat precarious. This is why languages die, knowledge fades, and skills disappear. Humans long ago developed ways to safeguard against such losses with storytelling traditions, informational hierarchies, and historical records in pictures and words. In large complex societies, where there are exponentially more mechanisms for information storage and transmission, there are a correspondingly greater potential number of connections between and among people.

The resulting patterns of bonds of functional interdependence make up the social context in which humans develop and live. Whether in a society of 50 or 50 million, how we meet our needs, what we learn, and how we become who we are necessarily takes place within the context of a web of relations through which necessary functions are served, needs and wants are met. These patterns, like the humans who comprise them, are dynamic—exhibiting both stability and change to varying degrees at different levels. Understanding this, we can begin to appreciate what the core concepts "individual" and "society" actually mean.

What We Get

Even if the above premises (and the knowledge that informs them) don't seem all that revolutionary, together they offer a subtle shift in perspective that is. For one thing, they remind us that human activities are seamlessly intertwined with, rather than in opposition to, biophysical processes. They likewise clarify the inherent sociality of individuals, in contrast with the usual individual-society antinomy. In both cases, refining our knowledge of the "starting conditions" referred to in the introduction, improves our chances of heading in the right direction. They also highlight the need for concepts which more accurately communicate those realities. To put it bluntly, as Elias did, we need to "give up thinking in terms of single, isolated substances and to start thinking in terms of relationships and functions" (Elias 1991:19). Changing our habits of thought in these ways cannot be achieved by sheer will. Rather, old thought patterns must gradually be replaced by new ones, a process that is aided by the use of more suitable concepts and mental pictures.

Upgraded Core Concepts

Most of us don't have a reason to think twice about the language we use every day, and so are fairly oblivious to its effects on our habits of thought. If we did, we might notice, as linguist Benjamin Whorf did, the implications of the way European languages construct sentences with a subject and predicate, as in "the river runs" or "the wind blows." This particular structure, Elias (1978) noted, implies a subject at rest, with change or action introduced through the addition of a verb or some other attribute—as if there could be an unmoving river or inert wind. This grammatical structure constrains our thinking in important ways, forcing us to assume a starting view of unchanging-ness before we can speak of change. In similar fashion, language influences our ability to conceptualize relationships and that which is related. The concepts presently available to us tend to portray interrelated things as separate and isolated, to which relationality must be tacked on. Without concepts capable of expressing inherent dynamism or relationality, we cannot clearly think or speak about phenomena which are inherently dynamic and relational. This helps explain the trouble we've had in trying to articulate the nature of human life in the world.

For those of us who only speak languages structured in this way, it can be difficult to imagine an alternative. But they exist, and they illustrate how different linguistic systems can facilitate different ways of thinking (Jia and Sun 2002; Jia and Jia 2008). Plant ecologist Robin Wall Kimmerer, for instance, writes eloquently about her ancestral Potawatomi as an example of a language predisposed to express dynamic relations. Composed primarily of verbs, it is highly capable of describing the vital beingness of the world. And, rather than using "it" to refer to nature—a practice that numbs us to the consequences of what we do—Potawatomi talks about all living beings as subjects. "Birds, bugs, and berries are spoken of with the same respectful grammar as humans are, as if we were all members of the same family. Because we are" (2017:17). Kimmerer argues not only that the dominant language is inadequate for comprehending what is actually taking place in the world, she advocates grammar as a tool of revolutionary change.

This makes sense. After all, one of the most distinctive powers of *Homo sapiens sapiens* (beings who know that they know) is the ability to observe and deliberately affect what and how they, that is, we, think. Because the concepts at our disposal influence both, we have the capacity to use new conceptual tools to change our habits of thought and therefore how we perceive, experience, and act in the world.

In particular, we need terms that can help us avoid some of the main pitfalls of modern Western social thought: the society-less individual, the self-existent society, both of them as mostly static entities which change only when acted on by some force, and the separation between humans and nature. Fortunately, alternatives that better capture dynamic relationality already exist. Before we get to talking about them, we'll consider a possible mental picture suggested by the above premises.

Mental Picture as Map

To briefly re-cap: As a distinct variety of social organisms, we are necessarily embedded in relations of interdependence within which we develop certain ways of seeing and operating in the world that feel quite natural, what we might call "second nature." In conjunction with a person's uniqueness, this socially shared way of being orients our activities—our perceptions, practices, and the products of them. As biological organisms, embodying and interacting with biophysical conditions, the cumulative expression of these activities has discernible impacts on the world and other people, thereby influencing the conditions within which configurations of human bonds of interdependence continue to develop. At any given moment, the above processes are happening simultaneously. And there is at the same time an overall order of conditioning influences at work (i.e., societies necessarily precede the development of a socially derived "second nature," which must first exist in order to be expressed, after which impacts follow—all of which is made of and happening in biophysical conditions). This universal pattern, shown in Fig. 4.10, is observable in any human society, at any scale.

Each component in the figure above, viewed from a high level of synthesis, represents enormous complexity, with numerous interacting parts and processes at multiple underlying layers. We can imagine zooming in on any of those, much like we can zoom in on a digital map to see greater detail at larger scales. For now, though, we'll use that same imaginative capacity to situate this in a larger continuum of time, where we can see the above pattern as just one "moment" of a series. To do this

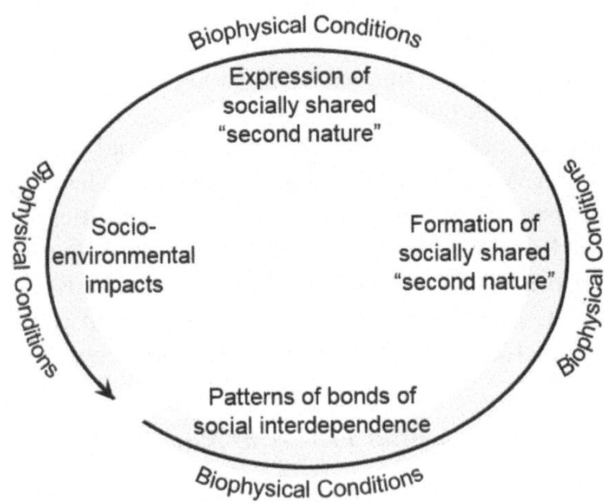

Fig. 4.10 Universal pattern of socio-environmental processes from a very high level of synthesis

we must first envision zooming out even further, to the point where we can see what physicists call the arrow of time (see Fig. 4.11).

Next, imagine moving in closer until you can see that what appeared to be a line actually has more than two dimensions. Figure 4.12 depicts a segment of that arrow, seen closer up.

Moving closer still, visualize this cylinder divided into slices, each one representing an undefined moment in time. Because time exhibits continuity, the slices are not separate (as in a loaf of bread) but resemble more

Fig. 4.11 Arrow of time

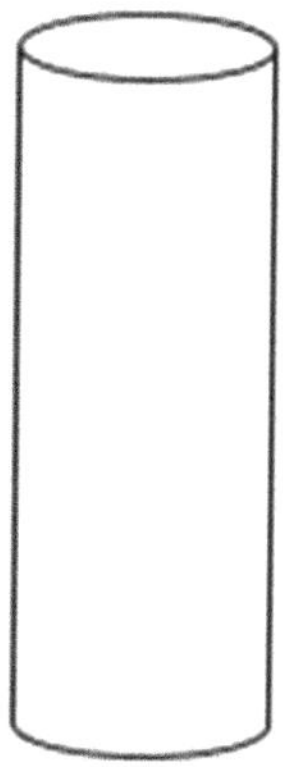

Fig. 4.12 Zooming in on a segment of time's arrow

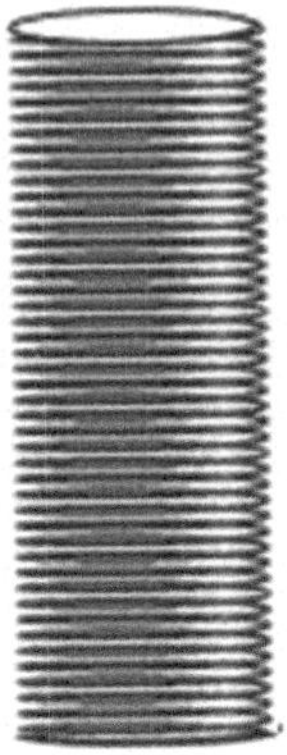

Fig. 4.13 Moments of time represented by layers of a spiral

the helical structure of a Slinky (the popular toy invented in the 1940s). The layers are connected in two ways: by virtue of the spiraling structure continually propelling the whole thing forward and the contact points they have with adjacent layers (see Fig. 4.13). We still can't make out any details from here, but it's a valuable reminder that what's happening at any given moment is necessarily conditioned by what came before and furnishes the basis of what comes next.

It's now time to re-visit the above statement that alternative concepts for referring to patterns of social interdependence and socially shared second nature already exist. Respectively, they are *figurations* and *habitus*. Elias is strongly associated with the first—a term he employed to refer both to the dynamic pattern of interdependent social relations as well as the very process of patterning. Regarding the second, he, along with many others, also made frequent use of the term habitus to denote the system of schemas that constitute one's socially learned way of being in the world. Together, as Elias used them, the correlated concepts of figuration and habitus serve as reminders that people are embodied interdependent processes. In this way, they help avoid the confusion that accompanies the usual opposition of individual and society. These concepts are considered in greater depth in Chapters 6 and 7 but are important to mention here as they, along with *biophysical conditions* and *socio-environmental impacts*, comprise the main features of the satellite picture shown in Fig. 4.14.

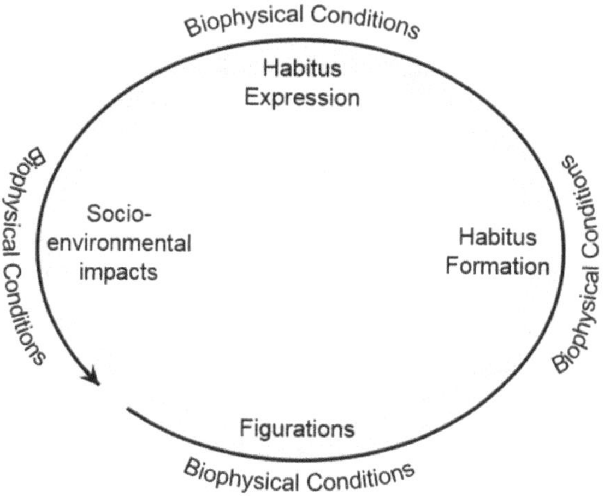

Fig. 4.14 Universal pattern of socio-environmental processes, using alternative concepts

Inserting this image at each moment within a larger continuum of time produces a simple map which can aid us in exploring long-term change, generating a fifth feature: *change over time*. What it lacks in detail, it makes up for in broader perspective, necessary for escaping the dead ends in which conventional concepts have kept us stuck. In the chapters that follow, we'll explore each of the model's components, and their relation to the whole, in greater depth. As we do, there are two important things to keep in mind. One is that the components, though conceptually separable, are in reality inseparable. Second, while they are ordered here according to the direction of conditioning influences, the phenomena they refer to are occurring simultaneously, albeit at different rates, on different levels, perceived from different vantage points. Beginning with the foundations of it all, we'll first turn our attention to biophysical conditions, tuning into how this particular category of conditions shapes our lives and how we, in turn, affect those conditions, thereby influencing the overall trajectory of social and environmental change.

Note

1. As a reminder, "small-scale" maps (somewhat non-intuitively) portray larger ranges from a greater distance and thus exclude much detail. Large-scale maps, on the other hand, show larger pictures of smaller areas in greater detail.

References

Black, Donald. 2000. "The Purification of Sociology." *Contemporary Sociology* 29:704–709.

Brotton, Jerry. 2012. *A History of the World in 12 Maps*. New York: Penguin.

Caldwell, Lynton. 1983. "Environmental Studies: Discipline or Metadiscipline?" *The Environmental Professional* 5(3–4):247–259.

Chen, Robert. 2013. "A Road Map Toward Better Understanding of Global Environmental Change." http://blogs.ei.columbia.edu/2013/03/04/a-road-map-towards-better-understanding-of-global-environmental-change/, March 4.

Commission on Physical Sciences, Mathematics, and Resources National Research Council. 1983. *Toward an International Geosphere-Biosphere Program*. Washington, DC: National Academy Press.

Doidge, Norman. 2007. *The Brain That Changes Itself : Stories of Personal Triumph From the Frontiers of Brain Science*. New York: Viking.

Edelman, Gerald, and Giulio Tononi. 2000. *A Universe of Consciousness: How Matter Becomes Imagination*. New York, NY: Basic Books.

Ehlers, E., and L. A. Kosinski. 1998. "From HDP to IHDP." *Global Environ Res* 1:95–96.

Elias, Norbert. 1991 [1987]. *The Society of Individuals*. New York and London: Continuum.

Elias, Norbert. 1987. "On Human Beings and Their Emotions: A Process Sociological Essay." *Theory, Culture and Society* 4(2–3):339–361.

Elias, Norbert. 1978 [1970]. *What is Sociology?* New York: Columbia University Press.

Fox, Sharon, Pat Levitt, and Charles Nelson. 2010. "How the Timing and Quality of Early Experiences Influence the Development of Brain Architecture." *Child Dev* January–February 81(1):28–40.

Ginsberg, Morris. 1937. "Sociology and Human Affairs." Pp. 166–180 in *Human Affairs: An Exposition of What Science Can Do For Man*, edited by R. B. Cattell, J. Cohen, and R. M. W. Travers. London: Macmillan.

Greenway, F. 1996. *Science International: A History of the International Council of Scientific Unions*. Cambridge, UK: Cambridge University Press.

Harley, J. B., and David Woodward. 1987. *History of Cartography*, Vol. 1. Chicago: University of Chicago Press.

International Geosphere-Biosphere Programme (IGBP). 1990. *The International Geosphere-Biosphere Programme: A Study of Global Change*. Stockholm.

Jia, Yuxin, and Jia Xuelai. 2008. "Revisiting Ancient Linguistic Worldview: East vs. West; Dao vs. Logos." *Intercultural Communication Studies* 17(4):77–96.

Jia, Yuxin, and Sun Benqing. 2002. "Contrastive Study of the Ancient Chinese and Western Linguistic Worldview." *Intercultural Communication Studies* 11(3):55–65.

Kates, Robert. 1985. "The Human Use of the Biosphere." Pp. 491–493 in *Global Change*, edited by T. C. Malone and J. G. Roederer. Cambridge: Cambridge University Press.

Kimmerer, Robin Wall. 2017. "Speaking of Nature: Finding Language that Affirms Our Kinship With the Natural World." Pp. 14–27 in *Orion Magazine*, March/April.

Lieberson, Stanley, and Freda Lynn. 2002 "Barking Up the Wrong Branch: Scientific Alternatives to the Current Model of Sociological Science." *Annual Review of Sociology* 28:1–19.

Lovelock, James. 1979. *Gaia: A New Look at Life on Earth*. Oxford, UK: Oxford University Press.

Massey, Douglas. 2005. *Strangers in a Strange Land: Humans in an Urbanizing World*. New York: W. W. Norton.

Miller, Roberta. 1989. "Human Dimensions of Global Environmental Change." Pp. 84–89 in *Global Change and Our Common Future: Papers from a Forum*, edited by R. DeFries and T. C. Malone. Washington, DC: National Academy Press.

Mooney, Harold, Anantha Duraiappahb, and Anne Larigauderiec. 2013. "Evolution of Natural and Social Science Interactions in Global Change Research Programs." *PNAS* 110(1):3665–3672.

Moran, Emilio. 2010. *Environmental Social Science: Human Environment Interactions and Sustainability*. Malden, MA: Wiley-Blackwell.

Pulver, S., N. Ulibarri, K. L. Sobocinski, S. M. Alexander, M. L. Johnson, P. F. McCord, and J. Dell'Angelo. 2018. "Frontiers in Socio-environmental Research: Components, Connections, Scale, And Context." *Ecology and Society* 23(3):23. https://doi.org/10.5751//ES-10280-230323.

Roederer, Juan G. 1988. "Tearing Down Disciplinary Barriers." *Astrophysics and Space Science* 144:659–667.

Rose, Deborah Bird, Thom van Dooren, Matthew Chrulew, Stuart Cooke, Matthew Kearnes, and Emily O'Gorman. 2012. "Thinking Through the Environment, Unsettling the Humanities." *Environmental Humanities* 1(1):1–5. https://doi.org/10.1215/22011919-3609940.

Smith, Kerri. 2009. "The Wisdom of Crowds: Climate Change Is Inherently a Social Problem—So Why Have Sociologists Been So Slow to Study It?" *Nature Reports Climate Change*. http://www.nature.com/climate/2009/0908/full/climate.2009.73.html; https://doi.org/10.1038/climate.2009.73.

Solbrig, O. 1985. "Chairmans's Summary: Life Systems." Pp. 221–227 in *Global Change*, edited by T. C. Malone and J. G. Roederer. Cambridge, UK: Cambridge University Press.

Stafford, Susan G., D. M. Bartels, S. Begay-Campbell, J. L. Bubier, J. C. Crittenden, S. L. Cutter, J. R. Delaney, et al. 2010. "Now Is the Time for Action: Transitions and Tipping Points in Complex Environmental Systems." *Environment: Science and Policy for Sustainable Development* 52 (1):38–45. https://doi.org/10.1080/00139150903481882.

Stock, Paul, and Rob J. F. Burton. 2011. "Defining Terms for Integrated (Multi-Inter-Trans-Disciplinary) Sustainability Research." *Sustainability* 3 (12):1090–1113. https://doi.org/10.3390/su3081090.

Trevors, J. T., and M. H. Saier. 2010. "Three Laws of Biology." *Water, Air, and Soil Pollution* 205(Suppl. 1):S87–S89. https://doi.org/10.1007/s11270-008-9925-3.

Victor, David. 2015. "Climate Change: Embed the Social Sciences in Climate Policy." *Nature* 520(7545):27–29. https://doi.org/10.1038/520027a.

Westley, Frances, Steven Carpenter, William Brock, C. S. Holling, and Lance Gunderson. 2002. "Why Systems of People and Nature are Not Just Social and Ecological Systems." Pp. 103–119 in *Panarchy: Understanding Transformations in Human and Natural Systems*, edited by Lance Gunderson and C. S. Holling. Washington, DC: Island Press.

Wilson, David Sloan, Steven C. Hayes, Anthony Biglan, and Dennis D. Embry. 2014. "Evolving the Future: Toward a Science of Intentional Change." *Behavioral and Brain Sciences* 37:395–460.

Zald, Mayer. 1991. "Sociology as a Discipline: Quasi-Science and Quasi-Humanities." *The American Sociologist* 22(3/4):165–187.

Zax, David. 2009. "The Last Experiment." *Seed Magazine*. http://seedmagazine.com.

CHAPTER 5

The Medium of Human Social Life

"We can never have enough of nature," Thoreau says in *Walden* (1899:333). This simple remark is literally true, but let's face it, a lot of people don't *feel* that way. In a culture where many have come to think of nature as something "out there," often associated with distant pristine places, many resonate more with Woody Allen's witty line: "I am at two with nature." The joke was original, but the sentiment is all too common. The mood of the modern age is characterized by this sense of alienation, a view of nature as other. Although largely taken for granted today, this view is particular to the modern age, as Elias has shown since his very earliest writings (2006).

Despite all the talk about reconnecting with nature, the sense of separation has grown. Children are spending less time outdoors than ever (estimates range from 4 to 30 minutes per day), with negative consequences for physical, mental, and emotional health (Louv 2005; Sampson 2015). Many of them become adults who spend little time outdoors, contributing to what has been called "the extinction of experience" with nature (Pyle 1993). Meanwhile, even concerted academic efforts to portray a more integrated relationship between humans and nature perpetuate our view of them as separate, as we saw in Chapter 4.

As they say, believing is seeing. Conceiving of humans and nature as separate limits our ability to see the dynamic interrelationships between

D. Kasper, *Beyond the Knowledge Crisis*,
Palgrave Studies on Norbert Elias,
https://doi.org/10.1007/978-3-030-48370-8_5

them and thus to act in accordance with those realities. As anthropologist Gregory Bateson put it, "the major problems in the world are the result of the difference between how nature works and the way people think" (2010). This premise should sound familiar; it's the inspiration behind campaigns for scientific literacy which, proponents argue, is crucial for enabling citizens to make wise choices in everyday life, discriminate between science-based conclusions and self-interested rationales, and evaluate the soundness of proposed policies (Priest 2013). Likewise, advocates of environmental literacy view "understanding and appreciating the natural world, and our place in it" as vital to the greater social good and global health, which is why it's been established as a fundamental educational goal (Environmental Literacy Council). It's generally agreed that "only an environmentally literate public will be able to find workable, evidence-based solutions for these challenges" (Hollweg et al. 2011).

Guided by this logic, we can think of our most pressing socio-environmental problems as symptoms of deeper issues with how we view nature and think about our place in it. It follows that meaningful lasting solutions must involve changes in our habits of thought, as Elias (2000) demonstrated long ago in observing the interdependent relationship between individual psychological structures, social relational structures, and long-term social change. Changing our habitual ways of thinking about the natural world and our place in it is no easy task given the concepts available and the challenge of seeing that which is everywhere, but it is eminently possible. This chapter aims to help in two main ways. One is to facilitate the capacity to see physical, biological, and social levels of phenomena as integrated, distinct but not separate. The other is to attune our awareness to the range of largely invisible conditions which (in concert with the mutual influence of human activities) have shaped, are shaping, and will shape individual lives, societies, and the trajectory of life on this planet. In what follows, we'll employ our new map and the concept of biophysical conditions as tools to help us cultivate the skills of noticing, navigating, and investigating the territory we're always in, but rarely paying attention to.

Biophysical Conditions

Encompassing the wide range of physical and biological phenomena which make up the world and all beings in it, "biophysical conditions" are literally the medium of human social life. As a concept for referring to the overall context of our lives, it has some distinct advantages over "nature" and "environment," terms more commonly used but which are saddled with implications of otherness, separateness, and further cultural baggage. Free of such connotations, "biophysical conditions" is a more objective and inclusive term for referring to the matter, energy, and processes which comprise the settings and stuff of our lives, and which we ourselves embody. Importantly, this includes the technologies and built environments that tend not to fit comfortably within the more conventional designators of nature and environment.

Figure 5.1 orients us on the map which, at a great remove, suggests a way to envision human social phenomena as emerging from physical and biological processes, rather than the usual way of seeing them set apart from and in opposition to "nature," "the environment," "ecosystems," and the like. The simple curved arrow and phenomena shown to emerge from it represent tremendous underlying depth, texture, and dynamism, which a simple exercise of mentally zooming in and out on those layers can evoke.

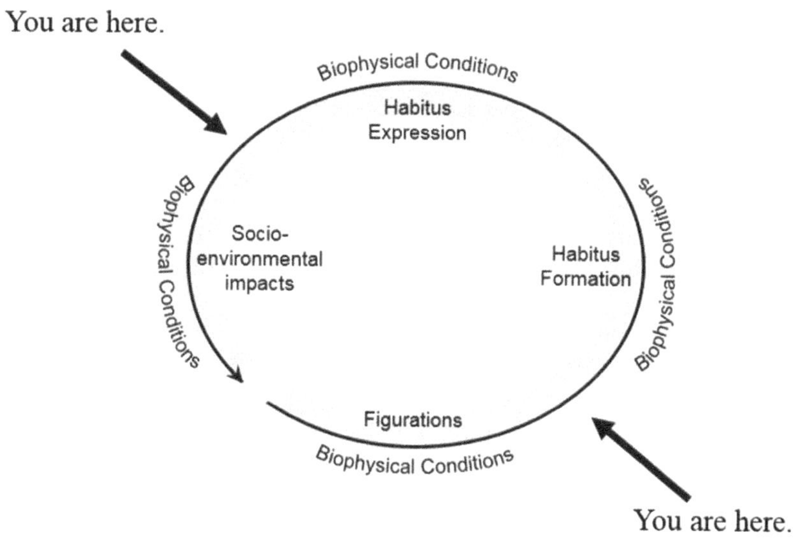

Fig. 5.1 Biophysical conditions. Locating ourselves in the model reminds us that biophysical conditions are the substrate of and pervade all human social processes

Just as we understand that life is made of yet distinct from "merely" physical phenomena, it's fairly easy to grasp the basic logic that human social processes are made of biophysical conditions, yet occupy their own distinct level. The trick is in being able to see (whether directly or with our mind's eye) the vast range of conditions, happening simultaneously and at multiple scales, in and with which we make our lives, form societies together, and exert our inevitable effects on others and the world. This is made even trickier by the fact that a good deal of these conditions is invisible to us. To render them "visible," we must employ established knowledge about biophysical conditions in combination with our powers of imagination. Below, we look at two main forms of invisibility and some of the strategies for seeing that which we normally can't or don't. The point is to consider how we can put those strategies to use in developing clearer ways of seeing and thinking about relevant biophysical conditions when we need to.

Stuff We Can't See

Whether too small, too vast, or too subtle for our senses to detect, there's a lot happening that we can't observe with the naked eye, or any other unaided sensory organ. Scientific advancements like microscopes, telescopes, space stations, and particle accelerators augment our observational capacity, but we still need to rely heavily on our imaginations to piece together what we learn and to see what existing technologies can't directly show us. Wielding available technologies and the knowledge derived from them with a great deal of creativity, scientists are able to enhance humans' already extraordinary capacity to envision what we cannot perceive on our own.

Using simple analogies, for instance, physicists translate ideas derived from complex mathematical equations and sophisticated experimental findings into pictures of the goings on at scales unavailable to our vision. In one such example, theoretical physicist Brian Greene instructs us to imagine standing on the ground and looking up at an ant on a suspended wire. From our point of view, it looks as though the ant has only two directions to go on the two-dimensional plane of this line (as in Fig. 5.2).

If we were to climb up and take a closer look, we'd see that there's more to the story from the ant's perspective. The ant is actually moving in three-dimensional space and can go back, forth, and around clockwise and counterclockwise (see Fig. 5.3).

Fig. 5.2 Watching an ant on a wire. What we see from a distance

Fig. 5.3 What the ant experiences

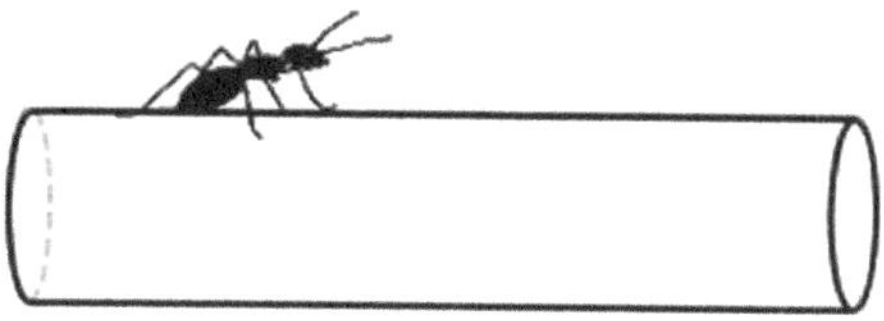

Greene's point is that, although we can easily perceive dimensions of a certain size, there may be others, in the microscopic depths of space which are curled up like the circular part of that wire, and so small we can't see them. Imagine a grid with small circles sticking up out of it, as in Fig. 5.4. "If you were a little ultra microscopic ant walking around," Greene (2005) explains in a TED talk, "you could walk in the big dimensions that we all know about—that's like the grid part—but you could also access the tiny curled-up dimension that's so small that we can't see it with the naked eye or even with any of our most refined equipment."

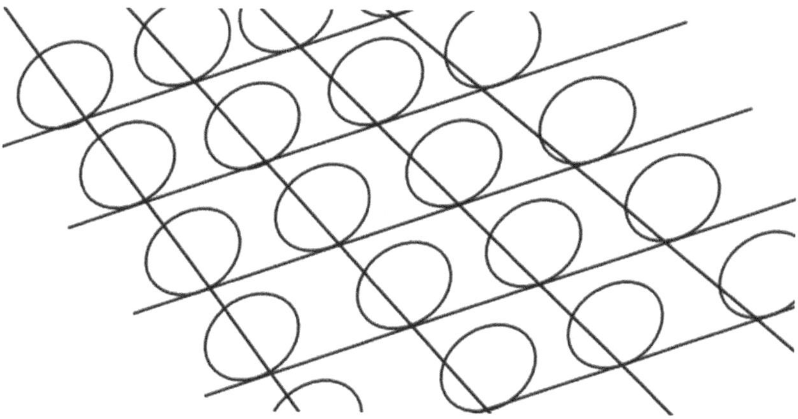

Fig. 5.4 Extra dimensions. The circles represent an additional spatial dimension curled up within every point of our familiar three-dimensional space

At the other end of the size spectrum, astronomers have devised ways to portray galaxies, superclusters, and giant superclusters—the largest entities known in the universe to date. In recent years, they've been busy mapping our own supercluster, *Laniakea*, of which the Milky Way is part. The magnificent images that work has generated offer an evidence-based way to envision the very large as never before possible. And, zooming out even further has revealed another new pattern in the universe, consisting of voids and densely packed superclusters (which, incidentally, rather resemble images of neural networks).[1]

Bringing things back down to earth, we find that many of the important material aspects of our lives—food, energy, and transportation systems, for instance—are similarly too massive and sprawling to see for ourselves all at once. Hidden infrastructures further obscure their visibility. For these reasons, the many kinds of diagrams and other visual aids developed to help us picture these complex systems have become invaluable. These and other examples remind us of the insights mental pictures afford. But what about the biophysical conditions we're *capable* of seeing but usually don't, or at least not very clearly?

Stuff We Don't See

As participants in a culture, people naturally take a lot for granted. The languages we speak and the values, beliefs, norms, and material contexts that direct our lives are just a few of the things that comprise what most grow up thinking is *just the way it is*. There are two main reasons for this.

For one thing, the ubiquity and enduring nature of much of the stuff of our lives causes it to resolve into an invisible background we don't much notice—until jolted, or guided, out of our oblivion. That's what the proverbial fish in water analogy is about. This mode of operating works fine for the fish, and it's usually okay for us too. After all, there are vast quantities of information coming at our sense fields at any given moment that the brain needs to prioritize. From its perspective, there's not much benefit to expending a lot of energy noticing what appear to be permanent and often benign fixtures of the world.

Another reason for our partial blindness is that we don't actually perceive the things of the world directly. Rather, our perceptions are mostly mediated by meanings, which limit and color what we (believe we) see. Through a lifelong process of social learning—some of which is explicitly taught, but much of which is tacit—we acquire associations, meanings, and ways of interpreting that our brains use to assemble stories about what we see. This form of mediated vision usually works well enough, but it's important to recognize that the stories our brains tell are not complete or accurate representations of reality.

In some ways, we already know this. The sun doesn't actually rise and set. Boys don't naturally have shorter hair than girls or an aversion to the color pink. Matter is not solid, as it appears to us to be, but is mostly made of space. This decade's clothing styles are not inherently more attractive than previous ones. Many of these stories are trivial (such as the perceived aesthetic appeal of fashion trends), while others are of great consequence (like the idea that women are naturally inferior to men). Some of the false stories we move beyond collectively (as with uniform lessons to school children that the Earth orbits the sun), while others are surpassed more gradually, in piecemeal fashion, by those fortunate enough to encounter information which has not yet become common knowledge (like the fact that biological race does not exist). Under normal circumstances, it's easy to take these stories to be solid and real, even though they are variable products of cultural conditioning.

There are correctives to both types of vision impairment. In the first case, it's simply a matter of looking. The difference between us and the fish is that we are *capable* of noticing "the water" and reflecting on our own minds if and when we want to. And there are reasons why, at times, it's crucial that we do. In the second case, while there's no physical lens that can show us the world beyond the meanings that accompany human perception, it is possible to develop our ability to see beyond our cultural conditioning. But it involves effort and is work that nobody else can do for us. With guidance, though, this work is made much easier and, with practice, the perspectival capacities it affords come to seem as natural as reading the words on this page.

This book is devoted to offering practical guidance in this respect. The current chapter focuses on helping us notice and better attend to biophysical conditions. Viewed along a continuum of changeability and informed by our simple hierarchical model of phenomenal reality, the concept becomes a powerful tool for seeing what's there *and* for strengthening our ability to see beyond conditioned interpretations.

Levels of Phenomena as Mental Shelving

It has become clear to me over years of teaching and learning that, unless one has a framework for organizing information—a place to put it, so to speak—it doesn't stay put. It's like having a closet without any shelves. With nowhere in particular to place things, we are left to simply toss items in and close the door. As the pile grows, so does the difficulty of finding what we need when we need it. Eventually it becomes uncontainable and, at some point, just opening the door causes stuff to spill out and fall away.

The challenge with learning about socio-environmental issues is that there's *a lot* to know. One can easily become overwhelmed. Environmental sciences and studies textbooks, attempting to pack a lot of key information into one place, do a great job of exposing students to the broad range of knowledge about physical and biological phenomena needed to understand complex multifaceted processes and problems. What tends to be missing is a reliable framework, the metaphorical mental shelving that gives readers (especially those new to the subject) a way to organize the vast quantities of information being thrown at them. Although there's a good deal of overlap in content, books are organized differently, each according to its own logic. As a result, the roughly two dozen chapters commonly covering subjects like Earth systems, evolution, ecosystems, sustainability, and the litany of usual problems (e.g., pollution, climate change, water quality and access, industrial agriculture, and so on) are experienced in rapid fire succession—a barrage of facts, data, scientific discoveries, and case studies.

To be sure, authors make earnest efforts to connect the dots and teachers, no doubt, try to be strategic in how they order the material. Still, in the absence of a comprehensive framework within which one can situate, appraise, and draw connections among available knowledge, one can easily become overwhelmed by the amount and diversity of information, much of which soon falls away.

On the other hand, the simple practice of distinguishing physical, biological, and social levels of phenomena and then employing those categories (like shelves) transforms the experience of encountering, consuming, managing, and using the vast quantities of information necessarily involved in doing socio-environmental studies. Most importantly, the nested nature of phenomena they depict serves as a constant reminder that *what's happening at the social level is inherently and always also physical and biological*.

Having an objective label like biophysical conditions for referring to the categories of phenomena that are always operating as the mutually influential context of human activities and the medium of our existence offers two distinct advantages. One is that it prompts us to notice that which surrounds and comprises us, allowing for a more seamless integration of social and biophysical phenomena in ways that conventional concepts do not. And two, it sets us up to discern important differences among those conditions, especially along a continuum of fixedness and changeability.

A Spectrum of Biophysical Conditions

Among the enormous variation in qualities of the biophysical conditions that comprise the medium of human social life, especially significant is the degree to which they are fixed or changeable. Observing this, we can make a necessary distinction between the conditions which are non-negotiable—those we must accept and work with—and those we can affect. This prepares us to more easily recognize the potential influences human activities are wielding (or could) on certain biophysical conditions and to consider why we may or may not want to do that.

Below, I highlight three broad types along this continuum of biophysical conditions: fixed, relatively fixed, and more changeable. Of course, any discussion of an inexhaustible subject like "biophysical conditions" will barely scratch the surface. But considering even just a few examples within each type brings significant benefits. It heightens our awareness of biophysical conditions (necessary for combating our culturally conditioned anthropocentrism), supports our ability to see them in a more nuanced way than we can see "nature" or "the environment," and ultimately facilitates a more reality-congruent mental picture of socio-environmental processes.

Fixed Biophysical Conditions

In reality, nothing is forever fixed. The sun will one day burn out. The universe expands or (and?) contracts. Even the reliably one-way direction of time physicists can technically only speak about in probabilities. But some time periods are long enough, some probabilities close enough to a sure thing, and some principles universal enough that we can treat them—for all intents and purposes—as if they are certain, unchanging, fixed. So that's what we'll call them. These "fixed biophysical conditions" (e.g., gravity, electromagnetism, time) structure our existence at all levels. To get a glimpse of *how*, let's look at a couple examples of fundamental fixed conditions. Whether the information presented below is elementary or somewhat new to readers, thinking about fixed biophysical conditions in the context of the socio-environmental synthesis framework sharpens our ability to perceive them when we want to and helps us appreciate what they mean, in practical terms, for humans today and our efforts to deal with urgent crises.

Matter and Energy

Everything is made up of energy and matter. Matter takes up space and has mass. Energy is what affects matter, causing changes to its position or state, and exists in two basic forms. Kinetic energy is in motion (e.g., light traveling in waves, the movement of particles as heat, the flow of an electric charge), while potential energy is tucked away in storage (e.g., in a stretched rubber band, a battery, the bonds of atoms and molecules) and released only under the right circumstances. It would be true to say that the concepts of matter and energy cover all phenomena in the universe, but that's not especially helpful for our purposes. In order to develop the abilities to perceive and understand the practical significance of these biophysical conditions for our lives, we need to get a little more specific.

The building blocks of matter are atoms, molecules, and the known elements which, organized in different ways, make up the materials of the world. Some forms of organization result in non-living materials, like metals and hydrocarbons, while others are assembled in more complex ways to produce life. The really astonishing thing is that all the complexity of living things on earth comes mostly from a handful of elements. About 95% of our bodies, for example, are made of just four types of atoms—carbon, hydrogen, oxygen, and nitrogen—which combine to form the great variety of complex molecules which make up the proteins, carbohydrates, lipids, and nucleic acids that make us what we are and which provide the nutrients necessary to sustain our lives. There are other kinds of matter our bodies need, but only in small quantities. We can't survive without sodium, for instance, but too much will kill us. And then there are elements that are downright toxic to humans, like lead or mercury.

So, the universe, the Earth, and our bodies are all made of matter. The thing about matter is that it can't be created or destroyed; it merely changes form. In compliance with this cosmic rule, atoms endlessly disassemble and reassemble to create something new, making the universe the ultimate recycler. What's behind all this movement? Energy is what causes changes in the position or state of matter. And energy has its own laws to follow, one of which dictates that energy, like matter, also can't be created or destroyed. It can only be converted from one form to another and, in the process, some useable energy is inevitably lost and disorder increased. The measure of this is called entropy.

Those not presently inclined toward an interest in science might be tempted to assume that these principles aren't relevant to them, but they should think again. Consider, for instance, some of the energy uses central to our lives: for heating buildings, cooling and cooking food and drinks, maintaining our bodies, transporting people and stuff, and powering the various other machines we use. From the molecular level of proteins, lipids, and polysaccharides being broken down into smaller parts that our cells can use for energy or building blocks, to the familiar human scale where we start our cars and step on the gas to transform chemical energy into motion, to the unfathomable quantity of energy raining down from the sun every minute—nothing happens without energy.

In fact, when we consider the multitude of ways we use energy, we find that virtually all of the energy we consume originates from the sun. Its constant radiation not only warms our bodies directly, it gets held in the Earth's atmosphere to keep life on this planet in the comfort zone. Then there's food. As the Book of Isaiah reminds, all flesh is grass. Sunlight is photosynthesized by plants to make food for themselves and their microbial partners in the soil. Those plants are then eaten by animals (which includes us) to meet their bodies' energy needs and vicariously consumed by those who eat the animals who eat the plants (also us). Humans have developed all kinds of other non-biologically essential uses of the sun's energy. We burn woody plants and trees (the growth of which was fueled by the sun) to access it in a more condensed form. More concentrated still are the by-products of decayed plant matter that have undergone millions of years of geological and chemical activity underground. Humans have been burning these sun-derived fossil fuels—in the form of coal, oil, and natural gas—in earnest for nearly 300 years.

This simple remembrance that energy is needed to affect matter enables us to comprehend the tightly bound relationship between the energy humans use and the degree to which we transform the stuff of the world into goods, and more people. It is commonly assumed that energy consumption grows *because* of higher population. That's not entirely wrong. More people using energy tends to mean more energy being used. However, because energy is a primary requirement for the transformation of matter into the form of a human body, we can also think of more energy as a precursor to more people, much like the availability of more calories is a prerequisite for consuming them. Figure 5.5 shows how closely population growth tracks rates of energy consumption.

But where did all this extra energy come from? Knowing that not all energy sources are created equal helps us understand the answer to that. Compared to other forms of energy humans use, fossil fuels have especially high energy density (the amount of energy stored per unit volume). That, along with their portability and initial abundance and ease of access, is what made them so popular. Knowing that, and seeing the breakdown of energy consumption by source, helps us understand where all this energy (and all of us people) came from. Working with roughly the

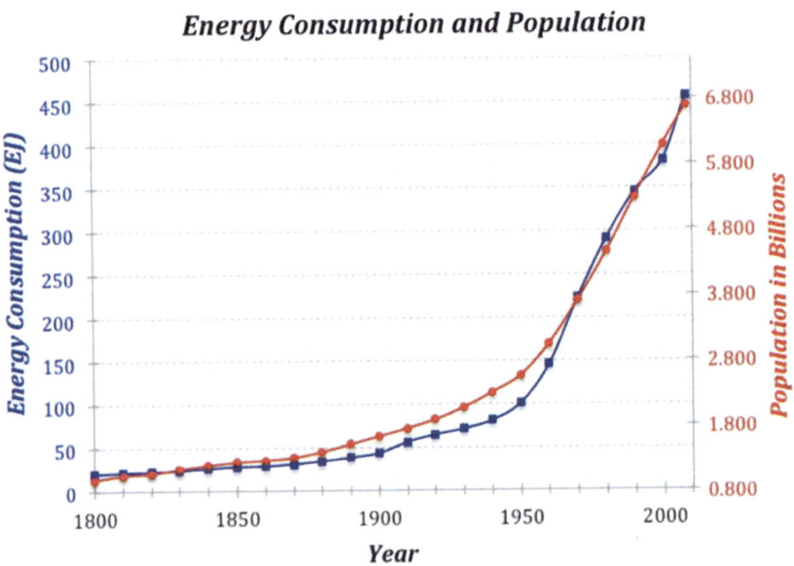

Fig. 5.5 Energy consumption and population. The relationship between energy consumption (shown as blue squares) and population growth (shown as red circles) over time (*Image credit* David Bice, with energy data from Vaclav Smil and U.N. population data)

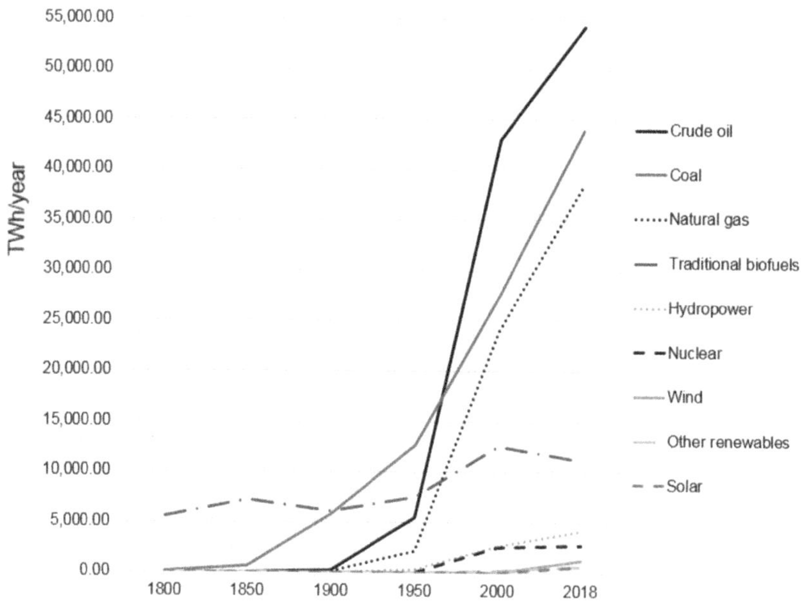

Fig. 5.6 Simple breakdown of global energy consumption by type, 1800–2018 (Data from Our World in Data, https://ourworldindata.org/)

same time scale as Fig. 5.5, Fig. 5.6 depicts the same increase in energy consumption—with rapid growth between 1900 and 1950 and very steep increases afterward—but breaks it down by energy source, allowing us to easily see the sharp rise in worldwide fossil fuel use. Our voracious consumption of these very dense forms of energy helps explain the exponential growth of humanity's energy usage and the population growth it made possible.

Universal Physical Laws

Ignorantia legis neminem excusat. This Latin adage, well-known among law scholars, stating that ignorance of law excuses no one applies well to the laws of physics. Nobody is exempt from them, regardless of what they do or don't understand. Considering just a couple of the most familiar ones, we can see all kinds of ways these non-negotiable rules play out in our everyday experiences.

Take the law of the conservation of matter. In our throw-away, flush-it-down, ship-it-off, burn-it-up culture, it's so easy to forget that the stuff that leaves our sight and our minds does not actually disappear. Ever. It's easy to conclude that the wood or coal or refuse we burn has vanished, leaving behind only a small pile of ash, but that would be wrong. Burning coal, for example, breaks the chemical bonds holding its carbon atoms in place and releases energy. The atoms and molecules comprising the coal do not go away, rather they are reconfigured through combustion-induced chemical reactions and released as air born pollutants such as mercury, sulfur dioxide, nitrogen oxides, carbon monoxide, and toxic heavy metals like lead and cadmium. Traveling by wind and water, they eventually settle in soils, concentrate in fats and blood streams, penetrate lung tissues, and do other harms in other places. At a larger scale, the relatively sudden burst of carbon dioxide caused by burning coal previously stored deep in the earth throws the Earth's carbon cycle out of balance. The resulting excess CO_2 in the atmosphere and oceans affects climate, weather, water chemistry and thus the fate of countless species. Actions have reactions. Causes have effects. It's as simple, and as complex, as that. And understanding the second law of thermodynamics as it relates to entropy helps us appreciate the importance of being attentive to those causes and effects.

Referring to the inevitable increase of disorderliness in systems, entropy is a measure of the amount of energy ***unavailable*** to do work. This principle has profound implications for basic human activities. Take light, for example. Contrasting an incandescent bulb—which converts only about 10% of the energy it uses to light and loses the other 90% as heat—with a LED bulb, in which the figures are reversed, we can clearly see that lost heat equals lost power. Likewise, anyone with an interest in distributing electricity across some distance should appreciate the significance of entropy, as they'll want to lose as little useable (and saleable) energy along the way as possible.

Being aware of entropy in our systems means accounting for the enormous amounts of energy it takes to maintain greater order in the system, maximize the energy available to do work, and capture what's been lost. Failure to account for all that can result in a huge net energy deficit, and the assorted consequences of our attempts to keep that credit line open. In a less technical way, this principle also reminds us that an ounce of prevention is worth a pound of cure. Applied to the problem of plastic pollution, where the plastic neatly embodied in a carelessly discarded bottle eventually breaks down into thousands of tiny bits dispersed far and wide, it highlights colossal differences between the efforts involved in, say, not producing or consuming single-use plastics in the first place, recycling all plastics produced and consumed, and gathering up the trillions of pieces of plastic (ranging from in-tact entire items to fragments to microplastics between 10 nanometers and 5 millimeters) now distributed throughout the world's waterways and oceans.

You don't have to be any kind of physicist or chemist to appreciate the significant takeaways here. Designing our lives and our societies in accordance with these cosmic commandments makes for systems which can be sustained and which can sustain us for a long, long time. Nothing we make is forever, but with cognizance of fixed biophysical conditions, we can create more sensible and satisfying human social systems that will continue into the foreseeable future. By contrast, systems which disregard the laws governing every aspect of our being inevitably create problems, some of which can ultimately bring about the demise of those systems.

Relatively Fixed Biophysical Conditions

It's all relative, we learned from one of the most famous physicists of all time. But when it comes to biophysical conditions, "relatively" fixed refers to the fact that, while these conditions are not *unchanging*, we can consider them fixed (to varying degrees) relative to the span of a human life or several generations of them. While the universal laws of physics really are non-negotiable, there are other conditions that *seem* unchanging but aren't. They feel that way because they change at a pace we can't easily perceive. It's important to be aware of these conditions so central to our experience—if, how, and when they might change, and whether and how we could potentially have a hand in that.

The Earth's Magnetism

Ever wonder why a compass needle points north when one faces that cardinal direction? Magnetism is the invisible force that moves it as the person holding the compass turns. Most of us live our lives oblivious to the fact that there's a giant magnetic field surrounding the Earth. The motion of convection currents of molten iron in the Earth's outer core, along with the spin of the planet, causes whirlpools of liquid metal to flow, generating electric currents that produce magnetic fields. The Earth's magnetic field in space (which extends thousands of miles, tens of thousands of kilometers) is called the magnetosphere. It protects the Earth from the charged particles of solar wind and cosmic rays that would otherwise strip away the upper atmosphere, including the protective layer of ozone that shields Earth from the sun's harmful ultraviolet radiation.

Since the invention of the compass in China more than 2000 years ago, most people have had no reason to question the stability of north, south, east, and west. We should know, however, that our geomagnetic poles are not permanently fixed; they periodically shift and reverse. As the title of one NASA article explains, "it happens all the (geologic) time" (NASA 2012). Evidence of these reversals is found in rock records, where the metals in the rocks are oriented in different directions depending on where the poles were during the epoch in which they were laid down. Based on this record, scientists tell us that reversals are the rule, not the exception. Apparently, Earth has settled into a pattern where pole reversal gradually, and then abruptly, occurs about every 200,000 to 300,000 years. Our north magnetic pole has crept northward by more than 600 miles (100 km) since the early nineteenth century (see Fig. 5.7). Since then, it's picked up the pace and is now migrating about 40 miles per year.

The magnetic field is changing in other ways too. Compass needles in Africa are drifting about one degree per decade, and the planet's overall magnetic field has weakened by about 10% since the nineteenth century (NASA 2003). Thankfully, even during times of transition, when the magnetic field gets more complicated during upheaval, it still protects us from space radiation and solar storms (see Fig. 5.8).

Shape Shifting Landforms

Somehow, my family ended up with two globes at home. One is older and shows the USSR and other former names of some of the world's nations. As someone who's interested in social change, I delight in these reminders that the human landscape, though it can seem inevitable and fixed, is not. But even when we're aware of how the social game pieces shift around on the board, we tend to think of the board itself—the masses of land on which people make a life and sometimes covet, invade, and fight over—as unmoving. Spoiler alert: they're not.

The giant-sized puzzle pieces we call continents are just what we can see at the surface. In reality, those majestic landforms rest on massive slabs of rock called tectonic plates. The Earth's hot core and the roiling liquid metal around it cause convection currents in the molten rock beneath

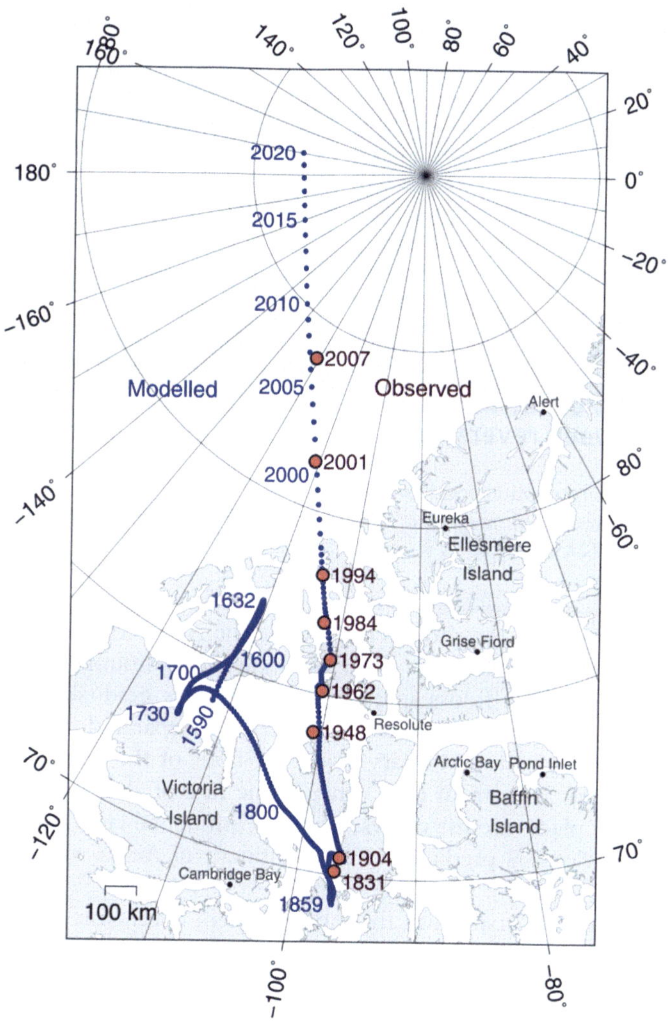

Fig. 5.7 The movement of Earth's north magnetic pole across the Canadian arctic, 1831–2001 (*Image credit* Geological Survey of Canada)

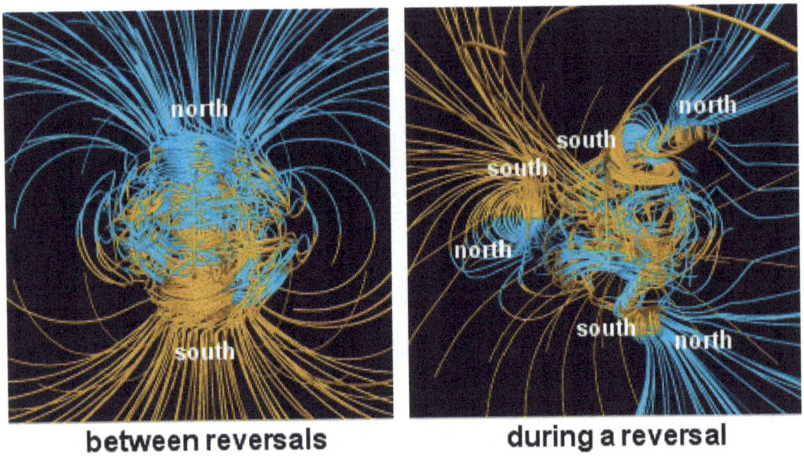

Fig. 5.8 Supercomputer models of Earth's magnetic field. Normal dipolar field (left) and complicated magnetic field during the upheaval of reversal (right) (*Image credit* NASA)

the Earth's crust. In the short-term, we see the results of this movement in earthquakes and volcanoes. Over the long-term, mountain ranges are raised and razed, sea floors spread, and continents gradually globetrot. North America and Eurasia, for example, are separated by the mid-Atlantic Ridge (a plate boundary along the floor of the Atlantic Ocean) and are moving away from each other at the rate of about one inch per year. In other places are rift valleys, where a continental land mass is being torn apart. Scientists predict that the continent of Africa will eventually split in two.

From a human perspective, the movements of continents happen over long periods of time, too slowly for those locational changes to affect us. Regardless, it's fascinating to consider the varying shapes our landforms have taken over the ages (see Fig. 5.9).

Even if we can't detect this movement with our ordinary sense faculties, the processes involved are of great consequence for us. Underneath its relatively calm exterior, with inputs of energy from the sun and the Earth's hot core, the world is in constant and intense motion. This energy not only moves the plates that slowly reconfigure our globe, it also moves the wind, water, and atmosphere in regularly patterned ways

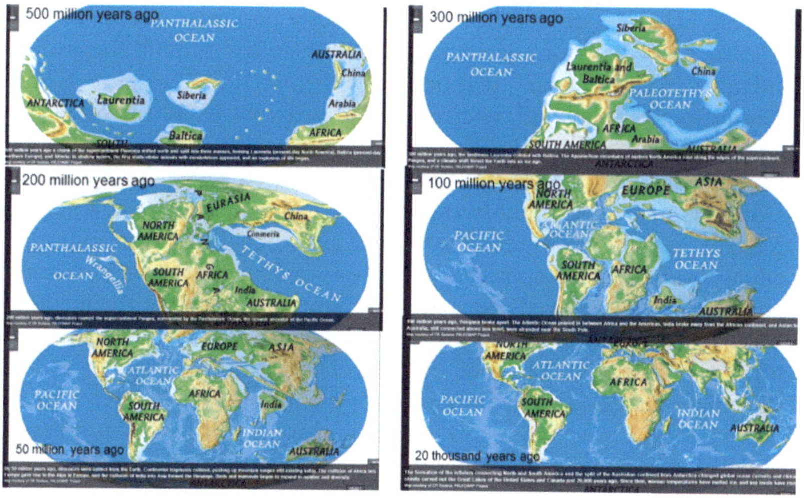

Fig. 5.9 The movement of land masses over millions of years (Adapted, with permission, from the work of Chris Scotese)

throughout the course of a year. These are phenomena we *can* observe in real time. More importantly, they make up the cumulative conditions to which *Homo sapiens* has adapted and on which we all depend. Speaking of, these evolutionary adaptive traits are another example of relatively fixed conditions with which, for best results, we must align human social activities.

Homo Sapiens *Characteristics*

Thanks to centuries of scientific labors, we know that species are not immutable; they emerge, change, come, and go over time in conjunction with environmental conditions and the element of chance. The smaller species are, the faster they can evolve. Whereas it took millions of years of evolution to produce the extant human species, *H. sapiens*, viral evolution can occur in a matter of weeks. Though scientists are beginning to be able to detect small, subtle genetic shifts that could contribute to superficial changes in the shorter term, humans have core traits that we can safely consider relatively fixed. As the result of eons of natural selection, our physiological and neurological conditions are the ground on which our mental, emotional, and social lives are built. These conditions not only affect what we do and how we do it, they permeate our very being and, as such, are worth understanding. And if we are to fulfill the ancient mandate to "know thyself," it is necessary that we do.

Earlier in the chapter, I referred to the proverbial "fish in water" to describe one type of human blind spot. However, it's important to note, a fish's life is more than just a matter of being *in* water. Its body, like our own, is made mostly *of* water and its anatomy is shaped largely *by* water. The water's density, chemistry, light absorbing capacities, and other qualities make up the biophysical conditions in which fish and all manner of aquatic life have evolved. To state that these conditions are the medium of a fish's life is not only to say that the fish exists within them, but also that the fish is largely a product *of* them. The same goes for us.

As biophysical conditions in particular environments allowed, natural selection worked over long periods of time shaping our bodies and brains in ways that were advantageous to human survival and reproduction. Some of those characteristics—like the fact that we stand upright or are diurnal—are *so* basic we scarcely notice them. If we stop to think about it, though, we can easily see how these qualities profoundly affect how we operate and set up our societies. There are several such non-negotiable organismal conditions that underlie, surround, and pervade what and who we are. Examples include our atmospheric requirements, stress response systems, nutritional needs, inherent sociality and the core motives it generates, and the functioning of the human brain (e.g., the energy requirements that drive it to economize, the capacities for logic and abstract thinking afforded by the neocortex, the subordination of rationality to the much older emotional brain, and so on). In later chapters, we'll dig into these qualities and how they shape our lives. For now, it's enough to call them to mind and remember that, on top of all these relatively fixed conditions, we must also contend with the more changeable biophysical conditions of the world, much of which *we* have a hand in shaping.

More Changeable Biophysical Conditions

While all biophysical conditions are subject to the laws and principles described above as fixed, we can identify a category of biological and physical conditions that not only *are* more changeable, but also that *do* change more rapidly relative to others. They span a wide range of temporal and spatial scales, from the changing scenery of human history to the moment-to-moment molecular and cellular changes happening in atmospheres and organisms. Among those, some are readily observable (e.g., landscapes, buildings, technologies), while others are perceptible only via observational and measurement tools (e.g., pollution, population characteristics, organismal processes). Together, they comprise the more transient dimension of the biophysical medium of human life. With the examples below, we'll practice the skill of noticing these conditions in our own lives.

Material Technologies

More often than not these days, the word technology is lazily deployed as a blanket reference to computers, mobile communication devices, and electronics, an unhelpfully narrow notion. Tech writer Kevin Kelly tries to correct this bias. By his definition, technology is anything produced by a mind. It includes a "domain of things on the planet that would only occur because of life but are, themselves, not living." Beaver dams and human dams, then, are both technology, as are spoons, levers, ropes, fireplaces, pencils, engines, and the endless assortment of things and processes our minds have produced and realized in the world. For some reason, people are fond of making proclamations about *the* most influential technology, but so many technologies are eligible for this honor that it's nearly impossible to pick a winner. In different contexts I've heard digging stick, writing, iron plow, machine screw, printing press, steam turbine, electric light, automobile, computer, Internet, and more offered up as *the* most world-changing tech. Adding to the challenge is the fact that no technology really stands alone. Most new technologies are the result of countless others that came before and were requisite to the new combinations and innovations that followed. Where does one draw the line? Thankfully, we don't need to pick a winner to see how a given technology has changed our lives and the world.

Technology occupies the category of biophysical conditions in two main ways. One is that it makes up the stuff of our lives. Looking around from where I sit in my small makeshift home office, for instance, I see technology everywhere: the computer I'm typing on, the lamp bringing light to this room in the pre-dawn darkness, all manner of writing utensils, stacks of books, the furniture that holds it all and me, the calendar that hangs on the wall, the mug that holds my drink, the space heater warming my feet in this cold corner of the house, the window through which sunlight can enter without inviting the wind and rain, and so on. Everywhere we turn is material evidence of technology; it makes up a huge portion of our physical world. More than that, our use of technology also *transforms that world*.

About 10,000 years ago agricultural technologies began to transform the landscape, claiming tracts of land for permanent cultivation and transforming societies and social life in countless ways. Since then, in the United States alone, certain agricultural practices have radically altered soil structures, eroded millions of tons of top soil, and turned hundreds of millions of acres of biodiverse forest and grassland ecosystems into long-term monocultures of commodity crops.

Another example is electricity. In addition to changing our homes, workplaces, public spaces and how we operate in them, the electrification of life has also spurred changes in our landscapes. The powerlines and high wires that border many roads and the underground cables stretched across vast distances connect to enormous booster stations and high voltage transmission lines, all of which have become part of the landscape of many people's lives. But there's more. Following the electricity this sprawling structure transmits back to its source leads us to a massive industrial power plant, the likes of which significantly change the biophysical characteristics of its locale. These plants operate on mined materials, the extraction and processing of which have additional devastating impacts on their places of origin. In sum, these technologies represent dramatic and far-reaching changes in the biophysical conditions of many people's, and other species', lives.

We can do this for any material technology: notice its direct presence in the landscape and then call to mind the other biophysical conditions accompanying it. Like an iceberg, the majority of the infrastructure underlying modern technologies is invisible, and requires some effort to see. Adding to the perceptual challenge is that, our acute awareness of new technologies which come on to the scene soon gives way to obliviousness, as they become part of the unseen backdrop of our lives, quietly transforming our visual landscapes, made environments, and even our bodies.

Bodies

Like everything else, *we* are made of matter and energy and are subject to the same laws that govern those phenomena. Bodies change in obvious ways over the course of a lifetime. This is clear in watching children grow, adults age, and witnessing the transformative effects of pregnancy, illness, stress, and exercise, for instance. During stretches of what appears to be relative stability, it's easy to forget that change is *always* happening in our bodies.

On average, a human body's cells are replaced every seven to ten years, with great variation in the longevity of different types. Skin, sperm, and intestinal wall cells, for example, live for days. Fat and some stem cells endure for years. Cells in the eye lens, egg, and heart muscle can live for decades. And experiments with neural transplants suggest that brain cells may be able to live beyond one's own life span and into another's. Throughout all this living, dying, and reproducing, cells are working busily on our behalf. Their cumulative activities result in the ongoing processes happening at tissue, organ, and system levels in our bodies. Oxygen is distributed and carbon dioxide hauled away as our breath and blood involuntarily circulate. Nutrients are broken down, transported, delivered, and transformed. Waste products are sorted out and eliminated. Hormones regulating essential bodily functions are secreted at intervals, fluctuating according to their own schedules of days, months, or years. Underlying our thought processes are neurons that recycle synaptic vesicles (tiny sacs filled with neurotransmitters that facilitate nerve impulses between neurons) in a tenth of a second and communicate with other neurons via electrical impulses about 200 times per second.

These are just a few of the long list of processes constantly happening in our bodies and those of all living organisms. Many of them are not yet understood. Thankfully, we don't need to know it all to grasp that, however unchanging a person appears to a casual observer at any given moment, that apparent stability coincides with constant change at many scales. Beyond awareness of the dynamic biophysical conditions that we *are*, it's crucial to appreciate the ways our inherent changeability manifests differently according to the confluence of conditions at the social level.

Science shows us, for instance, that air and water quality, diet, exercise, and social supports affect physiological systems, individual wellness, and the health of entire social systems. Likewise, the configuration of social systems and the habitus that develop within them can generate observable trends in certain biophysical qualities of people's bodies. The relationships between, for example, food systems and diseases like obesity, diabetes, heart disease, and cancer or between settlement patterns and weight are just a couple examples. In sum, our bodies themselves are changeable biophysical conditions which, in addition to and through normal biological processes, change in particular ways according to circumstances at the social level.

Upgraded Mental Models

Swapping out biophysical conditions for more conventional terms (e.g., environment, nature) in our mental models of socio-environmental processes may seem like a small and insignificant move. This small shift, however, especially accompanied by the ability to distinguish different categories of biophysical conditions, can make a huge difference in how we view the world and our place in it. And this is really the crux of the matter.

How we view nature, Elias presciently observed as a young man, "is intimately bound up with how we see ourselves and the world in general. It is, in short, very closely connected to all that makes up the peculiarity of a culture" (2006:10). Dislodging long entrenched ideas is no easy task, to be sure. But, as Elias reiterates in his post-script to *The Civilizing Process* 30 years after its initial publication, even if the ascendancy of more accurate images of the universe, the Earth, and human beings have not yet completely done away with the more ego-centered pictures and experiences, "it is certainly not impossible to dislodge this experience, and the image of man corresponding to it, from its self-evident acceptance" (2000:481). As he endeavored in that book, so too this work strives to promote:

> an image of human beings that agrees better with unhindered observation, and therefore facilitates access to problems which…remain more or less inaccessible from the standpoint of the old human self-image, or which, like the problem of the relation of individuals to society, continually give rise from that standpoint to unnecessarily complicated and never entirely convincing solutions. (Elias 2000:481)

It is in this spirit that I introduce the concept of biophysical conditions. Rather than starting from an assumption (whether explicit or tacit) of humans and nature as separate entities needing to be joined, the notion of biophysical conditions allows us to begin from a more accurate base of knowledge about the real conditions we're always working in and with.

In those conditions, physical and biological processes are never separate nor separable from social phenomena.

In addition, greater and keener awareness of biophysical conditions, properly understood, invites us to cultivate the capacities we naturally have for seeing the wide range of largely invisible conditions shaping our lives and societies. With explicit attention to those conditions which make up what we are, as well as the setting of our lives, the framework primes us to more clearly see the dynamism and contingencies inherent to human existence and the societies we form together. It gives us a way to think about what logic dictates, but which our concepts make difficult to comprehend and communicate: that human societies emerge *from* and are contingent *on* biophysical processes. Building on this knowledge, and using the map as a guide, we'll move on to get a closer look at human societies—those dynamic patterns of interdependent relations that form within particular biophysical conditions and are the contexts within which people, individually and collectively, develop and express particular ways of seeing and being in the world.

Note

1. See the video based on this work at: https://www.youtube.com/watch?v=rENyyRwxpHo.

References

Bateson, Gregory. 2010. Quoted in the Documentary, "an Ecology of the Mind," directed and produced by Nora Bateson.

Elias, Norbert. 2006 [1921]. "On Seeing in Nature." Pp. 5–22 in *Norbert Elias Early Writings*, translated by Edmund Jephcott. In *The Collected Works of Norbert Elias*, Vol. 1, edited by Richard Kilminster. Dublin: University College Dublin Press.

Elias, Norbert. 2000 [1939]. *The Civilizing Process: Sociogenetic and Psychogenetic Investigations*, translated by Edmund Jephcott. Malden, MA: Blackwell Publishing.

Environmental Literacy Council. "What Is Environmental Literacy?" https://enviroliteracy.org/home-page/about-us/what-is-environmental-literacy/. Accessed December 16, 2017.

Greene, Brian. 2005. "Making Sense of String Theory." TED Talk. https://www.ted.com/talks/brian_greene_on_string_theory.

Hollweg, K. S., Taylor, J. R., Bybee, R. W., Marcinkowski, T. J., McBeth, W. C., & Zoido, P. 2011. *Developing a Framework for Assessing Environmental Literacy*. Washington, DC: North American Association for Environmental Education. Available at http://www.naaee.net.

Louv, Richard. 2005. *Last Child in the Woods: Saving Our Children from Nature Deficit Disorder.* Chapel Hill, NC: Algonguin Books of Chapeil Hill.

NASA. 2003. "Earth's Inconstant Magnetic Field." December 29. https://science.nasa.gov/science-news/science-at-nasa/2003/29dec_magneticfield/.

NASA. 2012. "It Happens All the (Geologic) Time." November 30. https://www.nasa.gov/topics/earth/features/2012-poleReversal.html.

Priest, Susanna. 2013. "Critical Science Literacy: What Citizens and Journalists Need to Know to Make Sense of Science." *Bulletin of Science, Technology & Society* 33(5–6):138–145.

Pyle, Robert Michael. 1993. *The Thunder Tree: Lessons From an Urban Wildland.* Corvallis: Oregon State University Press.

Sampson, Scott. D. 2015. *How to Raise a Wild Child: The Art and Science of Falling in Love With Nature*. First Mariner Books.

Thoreau, Henry David. 1899. *Walden*. New York: T.Y. Crowell and Company.

CHAPTER 6

The Human Condition

The launch of the first satellite in 1957 represented, for some, a welcome step toward escaping our imprisonment to the Earth. Political philosopher Hannah Arendt responded to this sentiment, expressed in journalistic coverage of the event, by pointing out that, in fact, "the earth is the very quintessence of the human condition" (1958:2). Having just explored the biophysical conditions that are the medium of our lives, we are primed to appreciate the profound significance of that statement. We can now more easily recognize that the dynamic interdependence that constitutes and characterizes biophysical phenomena on our lively planet also constitutes and characterizes our relationships with other people. In this chapter, we'll explore dynamic interdependence as the fundamental and inescapable condition of human *social* life.

Despite the vague idea we have that we are somehow linked with others, our mental pictures of *how* we're connected and the terms we use to talk about that are exceptionally flawed. Similar to the way language systems can lead to the "it-ing" of nature (discussed in Chapter 4), modern western conventions in speech and thought reify our sense of society (and other social phenomena) as external things, obscuring our understanding of them. Fortunately, the reality of human interconnectedness is continually re-discovered, but we could save ourselves a lot of unnecessary labor by just remembering that. Embracing a social science built on concepts, theories, and models that more accurately reflect the

D. Kasper, *Beyond the Knowledge Crisis*,
Palgrave Studies on Norbert Elias,
https://doi.org/10.1007/978-3-030-48370-8_6

reality of human interdependence allows us to leave behind ways of thinking and speaking that keep us confused about the human condition.

Guided by Elias' (1978) clarity about the fact that individual and society do not occupy separate levels of social reality, but are distinguished simply by their reference to either interdependent persons in the plural or interdependent persons in the singular, this chapter and the next (its complement) explore alternative means for more accurately thinking and speaking about individuals as interdependent processes and societies as the dynamic patterns of interdependence within which they develop and operate. Just as situating humans in their biophysical contexts precludes the need to "re-unite" society with environment (solving all sorts of problems for socio-environmental studies), situating humans in dynamic patterns of social interdependence, and making this the conceptual starting point and object of sociological investigation, resolves longstanding problems in the social sciences. It enables us to finally dispense with fruitless debates about the relative importance of individual versus society, allows for more robust empirical study of human society, and ultimately provides the basis of a reliable theory of social change.

These are bold claims with significant implications for how we study human social life, but their logic is simple and easy to grasp. This chapter employs the concept of *figurations*, introduced and used consistently by Elias, to explore these claims and in the process demonstrates its value for orienting ourselves in and among mostly unseen webs of connection and for advancing sociological (and social) thought in the ways long called for (see Fig. 6.1).

Misperceiving Our Situation

To say that the individual is part of a larger whole formed with others is a very banal observation. Or as Elias pointed out, "it *would be* a banal observation if so many people did not constantly fail to register this simple state of affairs" (1991:11). Even though it's obvious that concepts like family, school, and nation refer to groupings of interdependent people, our habits of thought and speech turn them into external objects. My students demonstrate this tendency in a couple different ways when asked to sketch how they imagine their relations with others in society. Some depict themselves at the center of concentric circles of increasing size representing social entities larger than and outside of themselves, while

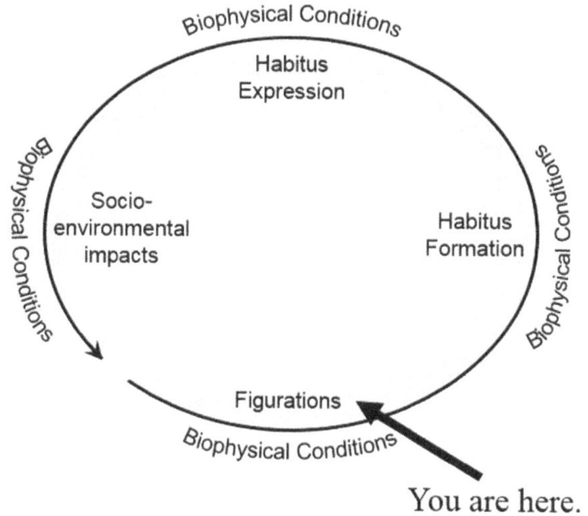

Fig. 6.1 Figurations. Locating dynamic patterns of bonds of functional interdependence (i.e., figurations) in the context of the overall processes taking place in a given moment reminds us that they develop within biophysical conditions, and that we develop within figurations

others portray social entities as discrete objects to which they are attached (see Figs. 6.2 and 6.3).

In the students' defense, this is a difficult exercise. For the most part, they've never been asked to imagine where they fit in the bigger picture, and it's not something they've been taught. If one had enough time and the right kind of prompts, one might eventually reach the point of drawing connections between themselves and the actual people to which they're linked. But what would that look like? How *should* we understand our place in society? As we learned in Chapter 3, not even sociologists have been able to agree on a standard definition of society, leaving us to intuit its meaning for ourselves using the terms at hand and prevalent cultural ideas as context clues.

The result has been a wide ranging set of interpretations, among which two contrasting tendencies have been especially prominent in modern western discourse. One is a view of society as self-existent entity external to and influential on individuals' lives, and the other is a belief that thinking, sensing individuals are the only entities that are *really* real and

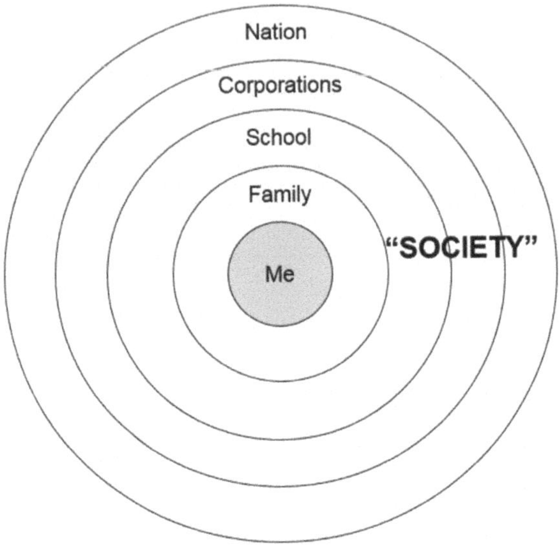

Fig. 6.2 Concentric circles view of self-in-society

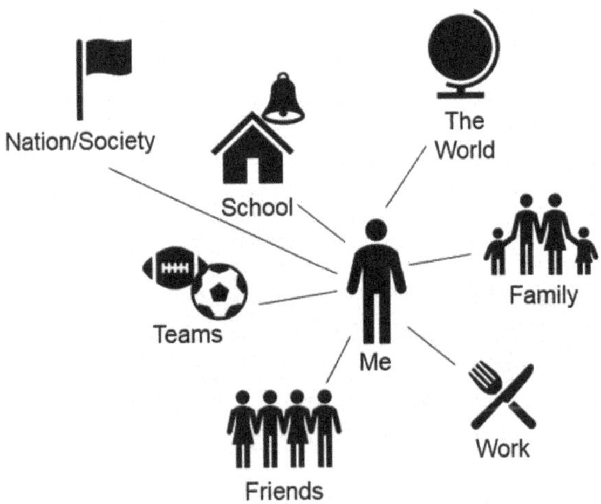

Fig. 6.3 Discrete entity view of self-in-society

have agency. These beliefs shape the ways we understand human action with important implications at individual and institutional levels for how we assess responsibility, attempt to elicit desired behaviors, and respond to others' actions. They also affect the course of history as tightly held ideologies about which matters more—individual or society—are played out on the world stage. The Cold War, for example, highlighted the clash of ideas about capitalist versus centrally planned economies. Within American politics are the dueling narratives of free market neo-liberalism set against social safety nets and societal accountability. Such conflicts are elaborate manifestations of the same basic modern tension between individualism and collectivism.

As divergent as these views seem, they have a common origin in erroneous assumptions about the autonomous existence of either individuals or society, or both. The problem is not so much that these instinctive notions are totally wrong, but that they are dangerously incomplete. Individuals *are* tangible, discrete, directly observable, choosing agents, whereas societies are not. At the same time, we *can* perceive the effects of larger social forces on our lives (e.g., recessions, fashion, slang, war) which are beyond any individual's control and we *can* trace many of our traits, tastes, and tendencies back to social sources. Each view reflects something true about social reality, but neither by itself is the whole story. Whatever form it takes, our misperception causes problems that go way beyond discrete interpersonal, political, and international conflicts.

Distortions in our view of social reality ultimately lend themselves to misguided visions of what constitutes a better world and how that translates into the goals and actions deemed necessary for realizing it. This is profoundly relevant for any efforts to generate change in a particular direction. As Elias states, "we can only clarify our actions, our goals and ideas of what ought to be, if we better understand what is" (1991:11). This logic is apparent in the unfolding of certain problems in the contemporary United States, which author and social critic James Howard Kunstler (2017) describes as the product of "a society that cannot form a coherent idea of what's happening to it, and as a result cannot make coherent plans for what to do about it." Ultimately, he concludes, the long-term results will be shaped by the degree to which we exercise our "capacity for meeting the exigencies of our time with ideas that comport with reality."

The stakes are high. Attaining a better understanding of "what is," as Elias phrased it, is more important than ever. One of the most essential qualities of "what is" that we need to understand is dynamic interdependence. We turn again to the biophysical realm for insights into what that really means.

Re-training Our Minds to See Dynamic Interdependence

During a class field trip to a Cleveland facility where some of the city's food waste is turned into compost, Nathan, one of the handful of employees who run the young enterprise, showed us around and then launched into a mini-lesson on soil. He began by asking students to name their favorite flower and describe how they would draw it. Following their instructions, we ended up with a sunflower on the whiteboard. Nathan then asked them what was missing from the picture? The students were stumped. Leaves, pistils, stamens? After some time of wondering, he sketched roots (murmurs of "oh yeah!" could be heard) and then added lots of shapes around them to represent the root zone, with its multitudes of organisms that together serve as the living foundation that keeps the plant alive. He drove the point home by saying that "the pictures of the world we have in our minds reflect how we think about the world" and that how we think about the world influences our actions. In other words, the more closely our thinking and mental pictures comport with reality, the more appropriate our actions will be.

Our mental image of a botanical entity as merely its aboveground parts ignores the other half of the plant from which its stability, survival, and growth derive via mutually beneficial relationships with multitudes of other beings. Likewise, our dominant view of a person as a discrete entity separate from the world of people and things beyond it is woefully incomplete. Even if we have some sense that "no man is an island," it tends to be vague, superficial, and fleeting. If we really want to understand what John Donne's iconic poem means, it's helpful to first recognize that, in fact, no *island* is really an island. As with the sunflower, when our eyes see a body of land isolated by water, we're only taking in a small part of the actual picture. The island we see is just the above-water protuberance of the continental shelf, volcanic crust, or ocean floor of which it's a part. This is true of many things in nature, including ourselves.

A person who appears to our modern eyes to be a self-contained unit is actually just the visible manifestation of a confluence of innumerable physical, biological, and social happenings. Contrary to popular ideas about the "self made man," the truth is that the totality of one's unique self comes into being *only* through relations with others. Just as we have done throughout history with so many physical and biological phenomena, we can train our minds to re-interpret and more accurately envision social phenomena. To do that, we need to break the habits of substantialist and dualistic thinking which imbue our ideas about individuals and society with a quality of independent self-existence.

A mere moment of careful thought reveals that, in reality, individual people are born, grow, change, and die in society with others and that every society is made up of an interconnected plurality of dynamic individuals. However, because of the concepts that presently structure our thinking about such matters, the ideas we naturally have of people as individuals *and* societies never really coalesce. As a result, the perceived gap separating these respective "things" persists. It is this gap which modern thinkers of all kinds have long struggled, and failed, to explain. Fortunately, the solution, as with most riddles, is surprisingly simple.

Solving the Individual-Society Riddle

"What goes on four feet in the morning, two feet at noon, and three feet in the evening?" demanded the Sphinx at Thebes in mythical Greece. The price for answering incorrectly was to be devoured on the spot. It's easy to imagine desperate attempts to get it right—positing different creatures, or beings that somehow magically transform from one thing to another. Both would have left the respondent dead wrong. Analogously, attempts to answer questions about human social life have long resorted to explanations of change that posit unobservable entities and mysterious links between individuals and societies through which transmission and transformation mysteriously occur, perpetuating our notion of social processes as "quasi-magical" (Bourdieu 1990:58).

The solution in both cases is quite simple. Oedipus got it right. It's a person, he explained, who crawls on all fours as a baby, walks on two feet through mid-life, and leans on a cane in old age. How'd he do it? When it comes to solving riddles, conventional thought patterns are usually what get in the way. Arriving at the correct response requires us to be able to think outside of those patterns. The result often feels like equal parts surprise and "duh, why didn't I think of that?" In the case of this ancient riddle, as with explanations of human social phenomena, the answer lies with a view grounded in empirical reality, from which we can plainly see that people are processes. Oedipus remembered that.

Likewise, an understanding of human social life as grounded in the reality of biophysical conditions dissolves the supposed mysteries of the individual-society relationship and provides answers to tired questions about: how these entities connect and affect one another, the mechanisms by which people internalize external social phenomena and externalize internal personal phenomena, which is the real causal agent, and which came first. Rather than invoking solutions requiring unknown or quasi-magical processes of linkage and transmission between imagined entities—or the alternative, which is to invent a whole new and overly complicated ontology—we simply begin with what we can easily observe.

People don't come into the world as fully formed adults. We are born of parents, grow up with other people, learn, develop, and age in a context of social relations. The individual person each of us becomes is a result of interactions among the expression of our genes in response to their (and our overall) environs, a unique stream of experiences, and the social contexts within which we develop and operate. "One can only gain a clear understanding of the relation of individual and society if one includes in it the perpetual growing up of individuals within a society" (Elias 1991:25). Remembering this, we know that, not only are we processes, we are *relational* processes. Dynamic interdependence is our undeniable and inescapable ground.

It's not that scholars haven't known all along that societies are dynamic and relational. Even sociologists on opposite ends of the methodological collectivism to methodological individualism spectrum concede that, at its heart, society is a collection of interactions (Martin 2009). It's just that the concepts available and their conventional meanings do not support the development and deepening of that understanding. As a result, countless pages have been written in the attempt to add relational texture to the flat picture of society-as-entity and to explain how change happens in supposedly stable systems, with little to show for the effort.

Sociology's basic premises (introduced in Chapter 4) are conducive to the cultivation of this understanding and, at the same time, weaken the habits of thought that obscure our vision. Humans are biological organisms and social by nature. Our exceptionally dynamic brains make us exceptionally changeable beings who not only can learn but must learn from others in order to survive and develop normally. As a result, we necessarily exist in relations of interdependence, the larger patterning of which makes up what we typically call "society" but misconstrue as a self-existent thing. Because the concepts at our disposal structure our perception and thinking, which shapes the actions through which we impact other beings and the world, adopting more adequate concepts is a crucial basis for being able to have greater control over those impacts. This is especially pertinent to addressing the socio-environmental crises of acute concern today.

A Helpful Alternative: Figuration

Critical of the epistemological individualism of Western philosophy and social science, which tends to reduce processes to static states and portray persons as "closed"—isolated minds locked inside separate "thinking statues"—Elias took a different approach. Like Oedipus, applying awareness of the underlying nature of human beings allowed him to avoid such blatantly wrong assumptions. "Starting instead from the premise of *homines aperti*—pluralities of interdependent human beings cooperating, competing or simply affecting each other in dynamic, interweaving figurations—Elias was able to avoid the reifying separation of mind and matter, individual and society and nature and nurture" (Quilley 2009:122).

Needing an alternative term to communicate this, Elias used *figuration* to refer to the dynamic pattern of interdependent social bonds humans form together *and* to the process of patterning itself. Its correlate, *habitus* (the focus of the next chapter), refers to the dynamic structure of accumulated social conditioning that humans necessarily develop and express. Together, these concepts serve as constant reminders that humans grow from children into adults through—*and only through*—relations with others, helping us "avoid the traditional quandary: 'Here the individual, there society'" (Elias 2009:3). They enable us to think of ourselves and our societies more accurately as interdependent processes, rather than discrete static entities. When we begin to perceive the world with minds re-trained in this manner, patterns of dynamic interdependence resolve into clearer view. Below, we will delve more deeply into how this concept serves as a helpful alternative, but first a word about the word.

A Note on Terminology

Elias acknowledged the need for restraint in introducing new concepts, wanting to avoid anything that would block possible channels of communication within and between disciplines. However, he saw sufficient reason to introduce a concept that could capture the relational and processual qualities of human social life. Though ambivalent about what to call the overall approach, Elias mostly settled on the word figuration to signify the pattern and formation process of interdependent bonds that humans form and re-form together.[1]

One advantage of the term is that relationality and process are implicit in it, unlike conventional concepts (e.g., society, structure, institution) to which they must be somehow added. Another plus is its relative unfamiliarity. Apart from the connotations of figurative art it might have for a few,[2] it has little conceptual baggage, allowing people to come to understand the concept on its own terms. At the same time, without intuitive associations, it might have a hard time catching on outside small academic circles.

When the subject comes up with American scholars one question often raised is: Why not just say network? On the one hand, network might work just fine. After all, social networks are defined as sets of "socially relevant nodes connected by one or more relations" (Marin and Wellman 2011:11).[3] As such they can aid the shift toward relational thinking. However, as currently understood and used in sociology, explains British sociologist Stephen Mennell (2017a), the word network is somewhat problematic.

> 'Network' has come to be used in connection with highly technical and mathematical network theory…And I think, too, that the term 'network' can be used to camouflage the failure…of nearly all American sociologists to be able to distinguish between 'interaction' and 'interdependence'. We are constrained and affected by the activities of far more people with whom we are interdependent at a distance than the people we interact with face to face.

In thinking of networks mostly as chains of interacting individuals, this approach ends up seeking "explanations of an interactionist kind—more psychologistic than sociologistic," he adds. The weaknesses of interaction, Mennell says elsewhere, fail to deal with the social reality of interdependence and the unequal balances of power it involves (Mennell 2017b:36).

I find these arguments persuasive, but there are additional reasons why the term network is not ideal.

Part of the charm of network analysis is that it seeks social explanations for social phenomena. In practice, though, it suffers from the same problems that most other contemporary theories do. In arguing that "causation is not located in the individual, but in the social structure," (Marin and Wellman 2011:13), network analysis ends up reinforcing a dualistic paradigm. Moreover, it has difficulty accounting for individual agency and culture (Snijders 2011:505), cannot adequately explain the mechanisms of how network structures affect actors (and other kinds of nodes), and is very limited in its ability to explain and facilitate an understanding of change. Even the area of "network dynamics" mostly accounts for changes in the presence or absence of relations between nodes at different points in time, but cannot explain them or why they matter. Whatever the shortcomings, it's likely that methodological developments in social network analysis will be invaluable for studying figurational patterns. For now, we can use far less technical means to visualize them.

Envisioning Figurations

If a picture is worth a thousand words, a moving picture is worth even more, which is why the metaphor of interweaving is helpful for envisioning the dynamic interdependence that figuration signifies. Comparing the social world to a tapestry in progress, with multiple participants as threads contributing to its warp and weft, we can imagine how each influences what's being produced without any one being able to control or perfectly predict the outcome. Threads end and some patterns cease; others come together to create entirely new patterns, suggesting a relational process that is emergent, contingent, and creative.

The metaphor works in other ways too. The connotation of fibers and threads as the base materials aligns nicely with our attention to the biophysical, alluding to ideas in physics about strings as the fundamental stuff of the universe. With regard to meaning at the social level, it fits with notions of the "thread of a story" or "spinning a yarn." Just as elaborate tapestries can contain symbols that tell the stories of a people, a big picture view of figurational interweaving allows us to trace threads and patterns back in the directions from whence they came to understand how things got to be the way they are and to project forward for insights about what's possible or likely. The artful words of fiction writer Kim Stanley Robinson help us see how a figurational perspective:

> makes us a thread in a tapestry that has unrolled for centuries before us, and will unroll for centuries after us. We're midway through the loom, that's the present, and what we do casts the thread in a particular direction, and the picture in the tapestry changes accordingly. (2003:667)

With an appreciation for how this concept can help us envision our dynamic interconnectedness with others, we're ready to explore the more literal aspects of figuration.

The Basis of the Human Condition

As beings whose very existence is the product of human interaction, who are born of and cared for by others, whose learning and development comes from others, and whose striving for gratification is in many ways directed toward others, interdependence *is* the human condition. Like threads in a tapestry, our selves are interlaced with others in countless ways. And it's no mere metaphor.

Elias was ahead of his time in stressing the inseparability of humans' biological development with social learning, but we no longer need to argue about the relative importance of nature "or" nurture. "Advances in neuroscience, molecular biology, and other fields now offer definitive *knowledge* about who we are and how we are cognitively wired" (Massey 2005:60). We not only know that the human species is "social to the core," we understand why (Fiske 2003:169). The human brain, built on the evolutionary foundation of the emotion-oriented and thus social relationship managing limbic system, forms the basis of cognition, perception, and reason. We know that a child's experiences in particular physical and social environments shape neural connection, pruning, and pathway building processes—culminating in a way of engaging with the world via a particular combination of language, knowledge, skills, perceptions, beliefs, implicit associations, practices, dispositions, and so forth. More recent research in neuroscience shows that we are quite literally connected with others at the neural level, which is the basis of our capacity for empathy and learning via imitation. As neuroscientist V. S. Ramachandran (2009) explained in a presentation, "there are whole chains of neurons around this room, talking to each other. And there is no real distinctiveness of your consciousness from somebody else's consciousness." The reason we feel a distinction between self and other is because of the receptors in our skin sending signals that counter the neural signals of connection. So, while we can in some ways experience another's pain, we don't perceive their exact suffering in our bodies and get confused about who we are. Experiments have shown, however, that when the effects of those nerves in our skin are taken out of the picture (through anesthesia, for instance) and we watch another person's hand being touched, we *do* feel it in our own hand. In short, our intra-neural networks are shaped largely by social interaction while we are at the same time inter-neurally linked with others—whether we realize it or not.

Likewise, at the scale of the whole person, we can see that people are intertwined through their dependencies on others to fulfill basic human needs. Contrary to conventional notions of functionalism in sociology, comprehending this requires no belief in self-existent external entities or a normative assessment of what's good for the whole. Rather, it's based on observable human needs and the many forms that efforts to satisfy them take. Borrowing a term from chemistry which refers to the measure of an atom's combining power, Elias uses the word *valence* to describe the qualities of outreach and openness that characterize humans' natural capacity for and inclination toward bonding. Although our own combining capacity is far greater than that of an atom (strictly limited by the number of electrons in its outermost shell), it's a useful term for thinking about how our valencies are engaged in the processes of fulfilling basic human needs and the wants that derive from them.

It is through these bonds of interdependence that humans end up serving particular functions for one another, giving rise to various formal and informal social roles and the collection of relationships that make up society. While these can take an infinite number of forms, Elias suggests three general categories of functions that humans universally serve for one another. Survival and development is a broad category including the various forms of care people require to survive, grow into, and thrive as more self-reliant mature adults, as well as the numerous goods and services procured via others throughout the lifespan (e.g., protection, food, clothing, construction, medicine, energy, governance, education, and so on). As core components of our biology, sexual and reproductive functions are significant behavioral drivers and comprise a second category. Finally, with a vast suite of emotional requirements and capacities as part of our evolutionarily adapted equipment, emotional needs compel humans to look to others for fulfillment and expression, "whether directly by face-to-face relationships or indirectly by their attachment to common symbols" (Elias 1978:137). In sum, figurations, representing much more than the mere "division of labor," encompass the totality of participants' dealings in their relationships with each other.

In many cases, these functional interdependencies overlap and are fulfilled through some of the same relationships. For example, we tend to be emotionally bonded to those with whom we are sexually involved and/or with whom we cooperate to care for children, as well as those who have a primary role in ensuring our survival and development. On a larger scale, the figurations which tend to knit people together for the common purpose of survival (i.e., they exercise control over the use of physical violence between members and with non-members, among other things), what Elias called "survival units," often serve as a point of convergence to which many individual emotional valencies are strongly bonded. Where people have been bound and integrated into units for attack and defense, this bond has been stressed above all others (Elias 1978:139).

With core sociological premises grounded in biophysical reality, it's easy to see functional interdependence as a natural aspect of multiperspectival human relationships. As such, it becomes clear that our fundamental connectedness with others is not something we can choose or forego. As social organisms, we are *by nature* inclined, able, and required to connect with other individuals and groups. Though the strength of individual proclivities to interact with others varies and the ways this basic capacity manifests are influenced by social and personal circumstance, social interdependence is nonetheless inherent to the human experience.

Figurations and the Sociological Sciences

The study of dynamic relationships (as opposed to static substances) requires "different types of theory, concepts and methods of investigation" (Elias 1997:360). In general, the more that subjects under investigation have characteristics of highly self-regulating systems and processes, the more system and process models are required for exploring and conveying regularities among parts and wholes (Elias 1956). Fostering an appreciation of the organizational complexity of social phenomena, a figurational approach prompts us to more carefully consider the question of which scientific devices are appropriate at which level. Devices useful in studying simpler units, such as laws or independent variables, may lose their function or become merely auxiliary when applied to more complexly organized phenomena. In the case of sociology, figurations are a more appropriate conceptual device and unit of study.

The concept of figuration "puts the problem of human interdependencies into the very heart of sociological theory" (Elias 1978:134) and helps us see how social structures and processes emerge out of the interweaving of the intentions and actions of many people, but in ways which are neither the result of any one person's will, nor a matter of mechanistic causes and effects. As such, figuration helps clarify sociology's main task, "to examine and explain such structures and processes of interweaving" (Elias 1997:360), and is thus the "proper object of sociological investigation" (Quilley and Loyal 2005:812). Furthermore, the focus on particular types of interdependence that the concept of figuration suggests provides a basis for more empirical and actionable sociological inquiry.

Being biologically other-directed in ways that serve the three categories of functions, people's bonds of interdependence form overall patterns. Though not directly visible, these linkages and the patterns they form are no less real or observable. The smaller the numbers of people involved, the more directly can figurations be seen; the more numerous, long, and dense the links and chains connecting people to one another, the more indirectly they must be investigated. It would be fairly straightforward, for example, to map figurational patterns within smaller survival units, like the foraging societies that populated most of human history (and likely with neighboring tribes, bands, or clans too). Here, one person's web of relationships likely includes everyone in the unit, though the exact configuration of a person's valencies will differ from that of others. Larger patterns of interdependence, in villages, cities, or modern nation-states can also be detected, but mostly indirectly through indicators representing those at higher levels of synthesis. Four such qualities are discussed below.[4]

1. Degree of functional differentiation

Functional differentiation involves the number of different roles that exist to fulfill human needs and wants in a given figuration. While it may not be possible to capture all of the nuanced informal roles people play for one another in satisfying functions related to survival and development, sex and reproduction, and emotionality, it is possible to devise a fairly comprehensive inventory of formal roles and many informal ones.

The number of occupations for which a society has distinct names is one example of a proxy for indicating degree of differentiation. Given the massive amounts of information at our disposal, data management capacity, and researchers' creativity, there are numerous possibilities for indicating functional differentiation.

2. Degree of functional integration

Webs of interdependence exhibit integrating as well as differentiating aspects. Not unlike the *holons* coordinating different functions at different levels in organisms (e.g., cellular, tissue, organ, system) human figurations also have nexuses through which functions are integrated and coordinated via parts and wholes at different hierarchical levels and horizontal clusters. With regard to survival units, this functional integration has taken many forms throughout history in leaders, councils of elders, monarchs and their royal court, federations, and so on. In modern times, the nation-state is typically one of the highest levels of integration seen. Though more recently, international bodies like the United Nations or other coalitions of nations represent further levels of integration. Among the range of simple possibilities for assessing degrees and processes of integration, Elias explains, one is to ascertain the number of hierarchically graded levels of integration found in the figurations being structurally analyzed. Similar numbers of levels indicate the likelihood of other structural similarities (1978:144).

Outlining a society's basic integrative levels and nexuses at given points brings our attention, in iterative fashion, back to the functions served in and by each level and nexus. In this way, we can attend to the (often neglected) relationship between processes of integration and differentiation, necessary for studying long-term social change. With process in mind, we can appreciate the fact that the dissolution and decay of levels of integration are happening alongside combination and development.

3. Power ratios among bonds of interdependence

Power plays a central role in human social life, but has not been easily accessible to social scientific study. In part, this is simply because the social phenomena to which the term power refers are complex. In the effort to simplify things, a single form of power (e.g., physical force or economic) is often taken as *the* source of power to which all other instances of exercising power can be traced. As a result, power has acquired some negative connotations and become suspect to many. Adding to the difficulty is the fact that, consistent with the conceptual problems discussed above, power is usually treated as a thing. Someone is most often described as "having" power, or not, as if it were an object to be possessed and wielded. One of the most important contributions of a figurational perspective is that it properly treats power, not as a thing which is inherently good or bad, but *as a structural characteristic of all human relationships*. Where there is functional interdependence among people, there are power differentials.

Reminding us that we depend on others and others depend on us, a figurational perspective fosters awareness that power is not a zero-sum game, but more of a ratio. Within the context of figurations, power ratios are a matter of the ways and degrees to which different parties depend on others and are able to compel or direct each other's moves. In so far as we are more dependent on or directed by others than they are on or by us (regardless of whether that dependency is based on use of force, our need to be cared for, our desire for money, healing, status, a career, excitement, or whatever), they have a higher proportion of power in that relationship. But one is never totally powerless. Even in a highly unequal power ratio, such as the relationship between parent and child, children exercise some power over their parents (according to the value parents associate with the child). A baby's behaviors, for example, at first instinctual and later more intentional, elicit particular actions from its caregivers. This simple example demonstrates the relational character of power and that, whether the power differentials are large or small, or the relationships simple or complex, "balances of power are always present wherever there is functional interdependence between people" (Elias 1978:74).

With respect to assessing power ratios in figurational research, there is a lot of room for methodological development. For now, suffice it to say that in smaller figurations, we can readily imagine devising ways of denoting the direction or degree of (in)equality in the power ratios between individual persons. Larger more complex figurations would likely require a more coarse approach, like assessing power ratios between *categories* of people and/or social positions (perhaps designated by functional differentiation and level of integration), rather than individual persons.

4. Rates of change in all of the above over time

Among figurational thinkers, it's understood that sociology's proper task is to investigate "long-term transformations in the relations of interdependence between individuals and groups" (Quilley and Loyal

2005:812). As such, we not only need to be able to characterize and perhaps quantify the qualities of differentiation, integration, and power ratios in figurations, we also need to attend to rates and degrees of change in them over time. Have these qualities remained stable over a given period? For how long? Have they changed? How, how much, beginning when, and how rapidly? What countertrends have been or are at work and to what extent have they influenced the dominant trends? These sorts of questions, along with assessment of the other key characteristics, comprise the beginnings of a more reliable and empirical approach to fulfilling sociology's proper task.

There is much work to be done in exploring the full potential of this approach. The main intention here is to point out the key qualities which make figurations operationalizable, in ways that other conceptual collectives are not, and to highlight some of the key areas in need of methodological development. There are already many sophisticated tools and technologies for gathering, processing, and mapping this kind of information. The point is that, in knowing what to look for, we are better equipped to select and make use of them.

Perhaps most importantly for the purposes of this book, in developing practical methods for studying "long-term transformations in relations of interdependence," we get a reliable way of understanding and studying social change over time—a development socio-environmental researchers have been clamoring for. We'll come to that subject in Chapter 9. For now, it's crucial to underscore the value of a figurational perspective by looking at some of the erroneous conclusions we draw without it.

With and Without the Concept of Figurations

Operating in a modern context of rapidly increasing numbers of people, ever-lengthening chains of functional interdependence, and greater reliance on central integrating hubs for the coordination of it all, people today frequently experience the effects of social forces beyond their control. Without concepts for making sense of the interdependence at work, it's easy to misconstrue the situation. One might assume, for example, that there must be a "someone" at the helm, leading us to overlocalize and personify these forces as individual persons, parties, or superhuman entities to which we can attribute the pushes and pulls perceived in our lives. We see this today in extremely partisan politics, conspiracy theories which assign blame or credit to particular leaders or teams of masterminds, and in the scapegoating of ethnic, national, religious, or other groups.

Another form of misinterpretation is apparent today in the development of an even stronger sense of individual independence, ironic in the face of increasing interdependence! The following comments made by a psychologist in a recent presentation on mindfulness are representative of this all too common error. "In our current modern times," he said, "we don't need to be connected to survive anymore." As evidence that human relationships are "not about my survival anymore," he cites the facts that we can get food delivered without having to see another human being and put alarm systems on our houses, precluding the need to have people around (Goldstein 2018). This notion is rampant in our high-tech culture. For those who live in places where a simple tap of the screen or a voice command can bring much of what we need (or want) to our doorsteps, this conclusion is understandable. It's also totally wrong.

Beneath the illusion of independence is actually a vastly greater degree of dependence on more people and processes than ever before in the history of humanity. Let's take a closer look at the takeout food that appears at our beckoning. Pretty much everything about it—from the ingredients in the food to the order and delivery process to the production and disposal of the packaging—implicates us in long and complicated chains of functions embodied in the roles and processes involved in producing it and getting it to our table. For most of us living in wealthy nations, the sheer ease of procuring our next meal conceals the massively complex web of interdependence that connects us with untold numbers of others, often at vast distances.

A list of some of the current categories would include farmworkers, researchers, developers, manufacturers, and distributors of seeds, chemicals, and farm equipment; processors, flavor and nutrition scientists, regulators, inspectors, policymakers, lobbyists, package makers, market analysts, advertisers, graphic designers, shippers, stores, restaurants, delivery persons; along with the corporations and their various employees, shareholders, lawyers, and others involved in these processes, and with the fellow consumers with whom we share a recognition of, brand loyalty for, and emotional connection with, a given product and its symbolic representations. As you can see, the meals we enjoy in such carefree fashion are actually part of a magnificently complex web. And this is not the end of the line.

Following any of the above threads a bit further, we'd find them inextricably intertwined with still more people involved in the systems and processes these modern functions have come to depend on: the various elements of the economic system through which commodity markets, investing, selling, and buying take place; the people involved in engineering and installing the technology and machinery used in so many of these processes (from growing and processing to storage and transportation to sale and preparation), the radio, newspaper, television, internet, and overall information superstructure on which advertising and other relevant communications depend. Underlying all of this is the sprawling energy infrastructure without which most of the industrial and global processes related to food production and consumption today would not be possible.

Now, contrast this with the act of eating in the context of smaller figurations. In a mid-sized foraging society of several dozen, using a radically different example to make the point clear, everybody would know everybody, and all of them would have some knowledge about and/or responsibility for harvesting plants, nuts, fruits, roots, fungi, and animals on an ongoing basis, as well as activities related to the processing, storage, and distribution of certain foods at certain times of the year. If we were to map the bonds of functional interdependence connecting people in this society to their food, the chains would be very short and relatively simple. Some would show a direct connection between people and food source; others would include a few additional links, with perhaps one or two integrative levels through which the getting, processing, and distributing of food is socially organized.

Moving forward in time, we can imagine a society in the early stages of transitioning to simple forms of agriculture and markets. Here we'd find additional linkages between consumers and growers and sellers. Though more complex than before, at a small scale it is still possible for people to directly see, or at least easily comprehend, the overall pattern of relations they depend on for food.

In contrast, because the chains connecting modern eaters to their food have become increasingly long, now wrapping around the globe, and the links so many and so intricately tangled, it's impossible for any one of us to see all of the connections from our own limited points of view. As a result, the all too human sources of the influences we may sense are invisible to us, and therefore mostly experienced as alien. Driven by an innate need to understand, but lacking a reliable framework for doing so, we devise ways to make sense of these phenomena—attributing blame or credit to some "other" or deluding ourselves into a false sense of autonomy. Food is just one example. This increasingly intense interdependence applies to almost all other aspects of contemporary human existence.

The resulting obscurity contributes to an overall sense of disorientation underlying much of the "alienation," "general malaise," "angst," "anxiety," and other maladies which have erupted in response to the strange new patterns of modern life. The encouraging conclusion here is that it *is* possible to develop a more accurate picture of these interweaving threads and thus to better orient ourselves within the social web that stretches far beyond any one of us. This alone will not resolve the psychological disease mentioned above, but it is at the very least a necessary part of being able to carry out sociology's task of making blind, uncontrolled processes more accessible to understanding and thus amenable to a greater degree of control.

The theoretical and empirical foundation that a figurational approach provides primes us in novel ways to attend to the connections among people and the overall patterns they form. Like Theseus using Ariadne's thread to find his way out of the Minotaur's labyrinth, we can follow these connecting threads backward and forward as far as we can (or care to) go in trying to discern figurational patterns and their formation and re-configuration over time. And the motivation for doing so is not just academic.

The Stakes Are Higher Than Ever

The sense of interconnectedness that Donne sensed and expressed so poetically in the seventeenth century has been felt ever more strongly in recent years and articulated many times over in diverse ways and for different reasons. After World War II, it became clear to some that:

> In a world whose people are becoming rapidly more interdependent and in which the external forces that control them are becoming more centralized, there is urgent demand for a rational basis for planning and responsible decision making. (Ford Foundation 1948 annual report, cited in Hammond 2003)

This process only intensified in the second half of the twentieth century, where "separate human societies in all parts of the world have become increasingly interdependent" (Elias 1997:359). More recent observers of this trend have noted the way it raises the stakes, especially when we account for ecological concerns. It is said that the essence of the current era is the "deepening interdependence binding humanity and Earth into a single community of fate" (Raskin 2016:ii). This growing interdependence makes it even more critical that we better comprehend the consequences of our actions, because they affect so many.

The proposed framework shows the impacts of human action largely to be the results of the collective expression of a way of being that develops within particular figurational conditions. Understanding those consequences and how to influence them, then, requires study of the beings doing the acting—*in their singular and plural forms*. Rather than two independently existing objects, individuals and the societies they comprise are "two different but inseparable levels of the human world," and our understanding of either level suffers unless both are constantly considered (Elias 1978:129). It is to the other level that we now turn.

Notes

1. He is known to have preferred the term "process sociology" over "figurational sociology," but the latter prevailed (Gabriel and Mennell 2011:20).
2. This refers to art that represents objects in the real world, in contrast with abstract art. It seems fitting that, as a sociological term, figuration is similarly intended to more faithfully represent the real social world, as opposed to abstractions of it.
3. A node refers to members of a network, usually persons, but in principle could include any units that can be connected to other units (e.g., websites, organizations, nations).
4. For an elaboration on these, see Elias (1978:128–145).

References

Arendt, Hannah. 1958. *The Human Condition*. Chicago: University of Chicago Press.

Bourdieu, Pierre. 1990 [1980]. *The Logic of Practice*, translated by Richard Nice. Stanford: Stanford University Press.

Elias, Norbert. 1956. "Problems of Involvement and Detachment." *The British Journal of Sociology* 7(3):226–252.

Elias, Norbert. 1978 [1970]. *What Is Sociology?* New York, NY: Columbia University Press.

Elias, Norbert. 1991 [1987]. *The Society of Individuals*. New York and London: Continuum.

Elias, Norbert. 1997 [1977]. "Towards a Theory of Social Processes," translated by Robert van Kricken and Eric Dunning. *The British Journal of Sociology* 48(3):355–383.

Elias, Norbert. 2009. "Figuration," in *Essays III: The Collected Works of Norbert Elias*, Vol 16, edited by Richard Kilminster and Stephen Mennell. Dublin: University College Dublin Press.

Fiske, Susan. 2003. *A Core Motives Approach to Social Psychology*. New York: Wiley.

Gabriel, Norman, and Stephen Mennell. 2011. "Handing Over the Torch: Intergenerational Processes in Figurational Sociology." Pp. 5–23 in *Norbert Elias and Figurational Research: Processual Thinking in Sociology*. Malden, MA: Wiley-Blackwell.

Goldstein, Elisha. 2018. "Your Environment Matters." Talk for The Awake Network's Mindfulness Online Summit, May 19.

Hammond, Debora. 2003. *The Science of Synthesis*. Boulder: University Press of Colorado.

Kunstler, James Howard. 2017. "We Are Living in a Moment of Unprecedented Incoherence." Interview with Rob Hopkins. https://www.robhopkins.net/2017/06/16/james-howard-kunstler-we-are-living-in-a-moment-of-unprecedented-incoherence/, June 16.

Marin, Alexandra, and Barry Wellman. 2011. "Social Network Analysis: An Introduction." Pp. 11–25 in *The Sage Handbook of Social Network Analysis*, edited by John Scott and Peter J. Carrington. Los Angeles: Sage.

Martin, John Levi. 2009. *Social Structures*. Princeton, NJ: Princeton University Press.

Massey, Douglas. 2005. *Strangers in a Strange Land: Humans in an Urbanizing World*. New York and London: W. W. Norton.

Mennell, Stephen. 2017a. From a personal email correspondence, August 1, 2017.

Mennell, Stephen. 2017b. "It's Just Me in the U.S." *Magazine of the British Sociological Association* (Autumn issue):36.

Quilley, Stephen. 2009. "The Land Ethic as an Ecological Civilizing Process: Aldo Leopold, Norbert Elias, and Environmental Philosophy." *Environmental Ethics* 31:115–134.

Quilley, Stephen, and Steven Loyal. 2005. "Eliasian Sociology as a 'Central Theory' for the Human Sciences." *Current Sociology* 53(5):807–828.

Ramachandran, V. S. 2009. "The Neurons That Shaped Civilization." Presentation at TEDIndia. https://www.ted.com/talks/vs_ramachandran_the_neurons_that_shaped_civilization/transcript?language=en.

Raskin, Paul. 2016. *Journey to Earthland*. Boston, MA: Tellus Institute.

Robinson, Kim Stanley. 2003. *The Years of Rice and Salt*. New York: Bantam Books.

Snijders, Tom. 2011. "Network Dynamics." Pp. 501–513 in *The Sage Handbook of Social Network Analysis*, edited by John Scott and Peter J. Carrington. Los Angeles: Sage.

CHAPTER 7

Second Nature

The idea that no two snowflakes are alike is familiar to most, but can be difficult to fathom, especially when in the midst of a landscape buried in the stuff. Anyone who's ever tried to catch and hold those falling flakes to get a glimpse of their one-of-a-kind-ness immediately discovers two challenges: they are too small to be able to make out enough detail and they very quickly melt into watery oblivion. Wilson A. Bentley, on the other hand, the man who first photographed individual snowflakes beginning in 1885, achieved magnifications of between 64 and 3600 times. The results were spectacular (Fig. 7.1).

Since then, new knowledge has complicated the story of the unique snowflake. For one thing, we now know, most snowflakes are actually irregular and misshapen. The snow crystals showcased in images like the one above are relatively rare—the supermodels of the snowflake world. Also, and more importantly, we have learned that every snowflake starts out more or less the same, each acquiring its singularity only through its passage to the ground. When a tiny droplet in the water vapor of clouds freezes onto a particle of dust, an ice crystal is created. Along the way, more water vapor freezes onto the primary crystal, building new crystals and ultimately forming the six arms of the snowflake. A snowflake's journey involves moving through a unique mix of temperature, humidity, velocity, turbulence, and other conditions affecting crystal formation.

D. Kasper, *Beyond the Knowledge Crisis*,
Palgrave Studies on Norbert Elias,
https://doi.org/10.1007/978-3-030-48370-8_7

Fig. 7.1 Snowflakes, photographed by Wilson Bentley, 1902. Public domain, Wikimedia Commons

Because no two snowflakes experience exactly the same combination of conditions, no two snowflakes end up with exactly with same crystal structure. The snowflake's uniqueness, it turns out, *is not given but made*, the result of the environments and conditions the raw material traversed on its way down. "Their different shapes," essayist Adam Gopnik explains, "are owed to their different lives" (2011). With these insights, we can appreciate the snowflake—both its reality and the story we tell about it—as a helpful analogy for how we humans see ourselves.

The modern age is characterized by an emphasis on individuality, a celebration of uniqueness, and a sense that it is our distinctive qualities which make us who we are. Looking at ourselves through snapshots frozen in time, we mostly miss the fact that who we are, much like the snowflake, is not given but made, and never by us alone. From a different vantage point, we can see that the self-contained unit we once imagined an individual to be is instead an interdependent process, and that what we conventionally call the "individual" cannot be understood without considering the social conditions within which it develops.

In this chapter, we look at some of the key ways we currently misperceive individual selves and the problems that ensue from that. The term *habitus* is offered as an alternative concept that supports our ability to think about and study persons as interdependent processes, contributing to the advancement of sociological and socio-environmental studies (see Fig. 7.2).

Just Who Do You Think You Are?

Self-esteem. Self-confidence. Self-help. Self-improvement. Self-fulfillment. Self-actualization. Self-love. Self-care. Inner self. True self. Be yourself. The ubiquity of phrases like these in contemporary American life lends itself to the impression that we are utterly obsessed with our selves. But who is our self? What is a self? How does one's self-identity relate to larger social groups? Answers to these questions are not as straightforward as we are often led to believe, and have varied wildly across place and time. With regard to that last question, history reveals fluidity in the balance between emphases on self and reference group or groups. Norbert Elias coined the phrase "We-I balance" to capture that fluidity and to remind us that both aspects are always present, even if one is sometimes weighted far more heavily.

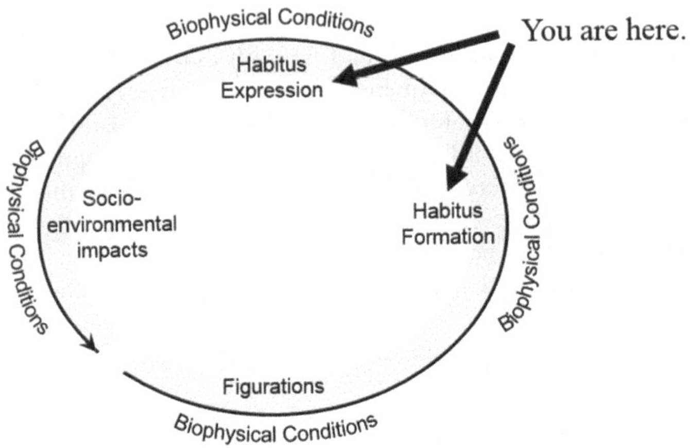

Fig. 7.2 Habitus. Orienting ourselves in the model, we are reminded that that habitus—the dynamic system of schemas which together generate a sense of "normal" or "second nature"—develops within figurations and its expression exerts impacts on other people and the world

A Shifting We-I Balance

Splitting the difference among estimates, evidence for full-fledged human societies dates back about 50,000 years. As far as we can tell, individualism as we know it did not exist throughout the vast majority of this time. As with any social species, cooperation and interdependence have always been humanity's insurance policy. Beyond mere survival, humans derived the roles, traditions, rules, and overall identities that shaped their lives *through* their place in larger patterns of relations. People recognized themselves and others as distinct individuals, of course, but that was just not as big a deal as it is today. More important by far were people's relations with the groups they were part of, from which they derived their We-identity. In our earliest foraging societies, the We-group was so fundamental to its members that to be banished from it was the harshest fate, the equivalent of a death sentence.

Only very recently do we see significant changes in the balance between the We-identity and the I-identity. As Elias' (1991) investigations show, ancient western languages did not have an equivalent concept for individual, much less one to designate society as its opposite. Though people clearly recognized distinctions among individuals, they apparently had no need for a universal concept which signified that. In medieval Latin, he explains, words like *individualis* or *individuus* referred to the indivisibility of something. As late as the seventeenth century, one could still speak of the "Holy individual Trinity," Elias tells us. Later, the word *individuum* came to be used to express a single case in a species of phenomena, a unique happening (e.g., *this* particular sugar maple, as a singular instance among many sugar maples). Recognition of the uniqueness of a single case developed in unforeseen ways. Eventually, the term came to be used mostly to describe the uniqueness of human beings.

Evidence from the histories of ancient states also shows the prominence of the We-identity, where the sense of belonging to a family, tribe, or state carried far more weight than any sense of an "I." This is reflected in the ways people labeled themselves, with the most important part of one's designator being that which indicated place of origin, clan ties, or social role (think Jesus of Nazareth, Elisabeth of Bohemia, Attila the Hun). In names consisting of both a personal and contextual name, the latter said much more about who a person was. The absence of last names up until less than a thousand years ago is further evidence of a more heavily weighted We-identity. And the fact that some groups continue to go without last names (like the Burmese, some South Indians, and Javanese) or place them before a personal name (as in Hungary, Hong Kong, Cambodia, China, Japan, Korea, Madagascar, Taiwan, Vietnam, and parts of India) reflects continued variability among the We-I balances of different cultures around the world.

I Think Therefore I Am

Although we continue to rely on social relations for many aspects of our survival and identities, how we *think* of ourselves and those relations has changed a great deal in recent history. Trying to reconstruct the development of this view in social thought, Elias looks to Descartes' and the circumstances that made a new form of self-consciousness possible. In brief, these involve the difficulties posed, in the European context, by the transition to a more secular worldview, a world in which long-standing religious authorities were gradually losing sway and in which individual thought and observation were gaining in the power to shape one's perceptions. Descartes' thinking reflects the growing awareness of his time that ordinary people could use their intellects to understand nature and put that understanding to practical use. One important consequence was the promotion and reification of "reason" in the human self-image. Eventually acquiring the appearance of an independent existence, reason seemed to emanate from a distinctly *other* being, a thinker, tucked away inside oneself.

Having trouble reconciling the different views of themselves as knower and known, people instead saw these as different components of themselves, made of separate substances and occupying separate planes of existence. The increasing ability to observe oneself and others in a more removed way "was consolidated into a permanent attitude and, thus fixed, generated in the observer an idea of himself as a detached being who existed independently of all others," an image Elias called *Homo clausus* (1991:106). This perception gradually hardened in thought and speech and gave rise to the sense that the real self existed as a separate entity *inside* one, cut off from the world by the walls of the bodily container, accessing external events only through the "windows" of the body's senses. Emerging trends in social thought reflected this, with philosophers becoming preoccupied with questions about how things and ideas "out there" get "inside" and endlessly puzzling over questions of how well our ideas about objects corresponded to the objects themselves.

At the same time, particular forms of social development were taking place that created a stronger sense of need to communicate about one's uniqueness. During the Renaissance, for instance, people in Europe's most developed countries were becoming more able to rise to relatively higher social positions. In the seventeenth century, we begin to see explicit distinctions, especially among the Puritans, between what is done individually and collectively. Later in the nineteenth century, this tendency was exacerbated by the growing social need for linguistic equivalents to antithetical sociopolitical movements and ideals, contributing to the more recent usage of "individual" and "society" as opposites.[1] Attention to historical fluctuations in the We-I balance and related ideas prepares us to appreciate the contingent nature of our own balance of individual to group identities and of the role that plays in our sense of self.

Seeing I's

In the United States today, we are steeped in a cultural ethos of individualism. From the "rugged individualism" that refers to one's supposed ability to make it on their own, to the notion of the "self-made man," to assumptions about the need to have one's own everything, our sense of individualism is reinforced by a massive rhetorical and social structure—evident in the notion of individual style, individually owned and operated communication stations, individual servings of individually packaged foods, individually targeted news and advertising, and more. The nation's, and increasingly the global, economy depends on sparking, stoking, and feeding the flames of the individual ego and its desires. Our politics tend toward favoring individual over collective rights and the culturally predominant mental habit is to attribute successes and failures to the merits of the individual (except sometimes, in cases of failure, when the individual is one's self). One doesn't have to be too astute an observer to see that we mostly think of ourselves as, well, individuals.

And we *feel* like individuals—like discrete and largely autonomous entities, obviously separated from others and the world by the visible boundary of our skin. At birth, we are assigned names and social security numbers intended to mark our unique identity. From a young age, we are encouraged to identify our personal possessions (e.g., toys, clothes, accessories, and spaces) as "mine." We are later encouraged to make a name for ourselves, to develop our own look, and to stand out from the crowd (though in ways that still allow us to fit in). In essence, we learn that our independence and the differences between our selves and others carry more weight than do our dependencies on and commonalities with one another. The fact that this seems so normal misleads us to assume that this is *just the way it is*. But as researchers who have studied self-interpretation across cultures know, even today, there remains a broad spectrum of how people understand their *selves*—ranging from individualist to collectivist (Triandis 1995) and independent to interdependent (Markus and Kitayama 1991; Vignoles et al. 2016).

As with our understanding of society (discussed in Chapter 6), how we understand individuality has implications for our ideas about morality, ideal social systems, and for how we view and interact with others. The implications of an extremely individualistic philosophy like Objectivism (developed by Ayn Rand and influential among some American libertarians and conservatives) illustrate the point. In this view, thinking, sensing individuals are seen to exist as metaphysically real objective facts, in contrast with social groups which do not. The logical conclusions drawn here are that: one's own happiness is the only proper moral purpose of a person's life, altruism is evil, any form of collectivism is to be rejected, and *laissez-faire* capitalism—the only social system consistent with this morality—is the ideal (Biddle 2014; Peikoff 2018). Most Americans don't take individualism *that* far. But we can confidently say that in the overall We-I balance today, the "I" weighs far more heavily, with a cascade of consequences for people and the social systems that affect their experiences and the course of their lives.

In sum, although the prominence of the individual appears in some contexts to be self-evident and like a universally valid way of seeing the world, a more expansive historical perspective shows it to be a relatively recent development running counter to much longer trends. Seeing how ideas about the concept of individual fluctuate can helpfully loosen our grip on the ideas we take to be true. This variability, though, should *not* be taken to indicate that there is *no* correct way to understand the self and therefore anything goes under the rule of cultural relativism. To the contrary, thoughtful observation and reasoning are sufficient resources from which to draw sound and accurate conclusions about the nature of the self, as is evident in the labors of religious adepts, artists, scientists, and thoughtful people like ourselves.

"Self" Is a Plural Verb

For 2500 years Buddhist thought has investigated and provided systematic training in understanding the nature of reality in which the things of the world, including the self, emerge from countless converging causes and conditions. Contrary to some initial misinterpretations of these teachings, they (at least in some schools of thought) do not imply that the self doesn't exist. Rather, they point out *how* the self takes place: not as a solid, fixed, independent, permanent entity, but as a continual confluence of dynamic conditions. Such insights are not restricted to eastern religions or exotic philosophies. In western literature, they have been expressed again and again by those who ask: "What shall we call our 'self'? Where does it begin? Where does it end? It overflows into everything that belongs to us—and then it flows back again"[2] and who observe that "a self is not something static, tied up in a pretty parcel and handed to the child, finished and complete. A self is always becoming."[3] More recently, scientific investigations corroborate these conclusions with empirical evidence, enhancing our understanding of the interdependent processes, at all levels, that characterize our selves. Below, we turn to two areas of study which are reinforcing our knowledge of that and enriching our abilities to envision how the self "is not a singular noun, but rather is a plural verb" (Siegel 2012:209).

The Me-Cosystem

In the realm of microbiology, it has become clear that the bodies we hold so dear are actually more *not us* than us. Each human is host to legions of bacteria, viruses, archaea, protists, and fungi. They inhabit our skin, guts, nasal passages, hair, the surfaces of our eyes, the enamel of our teeth, and pretty much everywhere else. Bacteria alone contribute about 100 trillion cells to "our" bodies, each of which is comprised of approximately 37 trillion human cells. Upon first hearing this, some react with a "*eeeewwww, gross*," an impulse that stems from the habit of thinking of our body as a separate self-contained entity. From this perspective, we see these creatures as invaders, to be ejected at all costs.

Beginning in the past century, we've acted on that view by waging warfare on the tiny beings, aggressively using anti-bacterial this and anti-microbial that on our bodies and in our homes, and demanding antibiotics at the first sign of a sore throat and sniffle. The narrative in the medical world has been: germs make us sick, so let's get rid of them as fast as we can. But we now know, explains microbiologist Martin J. Blaser, that "without most of these organisms we could never survive" (Specter 2012). Thanks to a growing body of research and half a century of the accumulated effects of our actions, we are discovering the costs of being such ungracious hosts to these small wonders and are coming to understand the critical, and mostly positive, role they play in human health.

For the past 150 million years, nearly all mammals have acquired their initial microbiome via passage through their mother's birth canal, which is colonized by an enormous range of bacterial species that seed the microbiomes of the skin and oral and nasal passages. The specific kinds of bacteria acquired there influence the way the rest of the body obtains the differentiated microbiomes we eventually end up with, a succession that is important for establishing a working gut microbiota, proper interaction with the immune system, and adequate early nutrition (DeSalle and Perkins 2015). The next sources of microbes, traditionally, are breast milk and the environments a child comes into contact with, including different ecological biomes, soils, animals, homes, and so forth. Later, the foods one eats shape the composition of microbes in our digestive tract, the most richly diverse habitat in our bodily ecosystem.

When we eat we're not just feeding ourselves. We're feeding our microbiome, allowing it to do its work of digesting our food, assimilating critical nutrients, manufacturing vitamins, bolstering our immune system, preventing infections, helping regulate brain chemistry, and much, much more (Montgomergy and Bikle 2016; Specter 2012). If we don't eat the foods the respective microbes need, they can't survive, much less optimally perform these critical tasks. We additionally compromise our wellness by eating that which feeds harmful microbes and engaging in practices that diminish microbial diversity—which is exactly what we're doing on a mass scale.

Researchers have discovered radical and recent changes in the microbiomes of Americans, especially in terms of reduced biodiversity. Investigations into the causes of these abrupt changes at the level of the microbiome are underway, but they correspond, and are thought to be associated, with significant social and behavioral changes during the same time period, such as: higher cesarean rates, lower rates of breastfeeding, excessive usage of antibiotics in children, consumption of products from animals fed antibiotics, a less diverse diet, increased consumption of processed foods, the ubiquity of anti-bacterial wipes and hand sanitizers, less time spent outdoors, and decreasing diversity in soil microbes. And the recent and swift rise of certain ailments—obesity, celiac disease, asthma, allergy syndromes, autoimmune disorders, diabetes, mood, and behavioral issues, for example—is increasingly being associated with deficiencies in the microbiome (DeSalle and Perkins 2015; Reid and Greene 2013; Specter 2012).

This fascinating and growing body of research is transforming how we think about wellness and illness. Replacing the notion of biophysical body as self-contained unit with an understanding of it as a highly dynamic and complex system of relations, health is no longer primarily a matter of exterminating invaders and exerting control, but is more about cultivating the *right relations* with the beings that support resilience in the body's systems. More than that, it also affirms the traditional wisdom that we are not who we think we are. From a biological standpoint, our self, rather than a single solitary being, is more like "an entire planet with a rich palette of ecosystems, as different as the Serengeti and Siberia, each hosting multitudes of microbes" (Montgomery and Bikle 2016:126). And those systems are intricately interconnected. In fact, we now understand, the microbiome is inextricably linked with neurodevelopment and mental as well as physical health. Tracing this connection, we move from what's happening in the gut to what's happening in the skull, where we find that even our capacity for thought is a wholly relational process.

It's Not All in Your Head

Looking back at historical oscillations in ideas about the brain, the mind, and where the true self resides—from the little man in our head to assumptions about blank slates or fixed mental hardware at birth to more sophisticated notions of mind as not just the brain but what the brain does—shows movement toward an appreciation of the thoroughly relational and dynamic nature of mental phenomena. The premises laid out in Chapter 4 remind us that the exceptionally dynamic human brain is the basis of our capacity and need to learn from others, making our brains inherently relational—both in the sense of relations among neurons *and* relations with others in the world. One consequence of this evolutionary twist, we now understand, is the perpetually open nature of the self.

The human brain is comprised of about 100 billion or so neurons, cells which receive, process, and transmit information through electrical and chemical signals. Though there are dozens of different types of neurons, they all exhibit the same basic structure. A single axon extends from the soma (cell body) like the trunk of a tree, from which numerous dendrites branch out (see Fig. 7.3). Amazing as they are, neurons are not much good by themselves. The real action occurs in relationships among them through points of connection called synapses. Signals are sent through the axon and received by the dendrites of linking neurons via synapses.

Though researchers don't fully understand exactly how all this chemical and electrical activity translates into awareness, emotions, concepts,

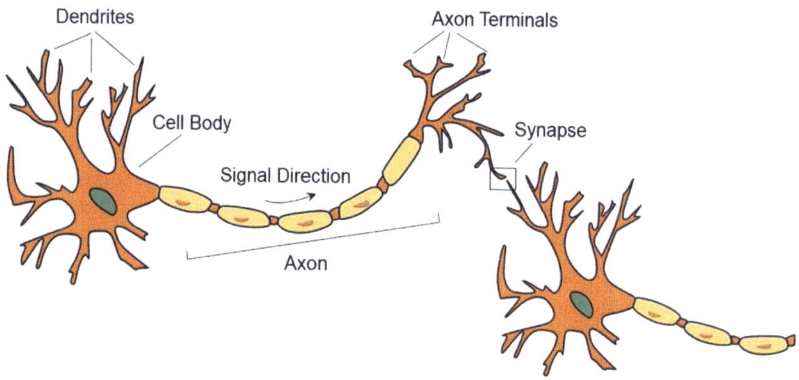

Fig. 7.3 Diagram of neurons and synaptic connection

thoughts, memories, and more, what they do know is that dynamic relations are at the heart of it all. Among the 30 billion neurons in the cerebral cortex alone, there are one million billion synapses, connection points between individual neurons. One pair of scientists put the magnitude of this number in perspective, calculating that if we counted one synapse per second, we would not finish counting for 32 million years (Edelman and Tononi 2000:38)! But it's not just the connections that are important. Also crucial in a person's development are *when* they form and within what *conditions*.

Just weeks after conception, the first neurons and synapses begin to form, a process that continues intensively throughout a child's crucial early years. Through the interaction of a baby's genes, environment, and experiences—including touch, talk, sight, sound, and especially reciprocal exchanges with others[4]—neural connections proliferate with astonishing speed: more than one million per second.[5] These connections build the brain's architecture, forming the foundation for later learning, behavior, and health. This initial burst of connectivity, which takes synapses to twice their neonatal levels by mid- to late-childhood, sets us up with widest range of learning possibilities. By adolescence, they decrease through a process called synaptic pruning. Just as one might prune an apple tree to get fewer, but larger, apples and stronger supporting branches, synaptic pruning is the brain's way of prioritizing quality over quantity.

Responding to cues about the kinds of learning needed (based on the frequency of certain stimuli and repeated use of different neural circuits which specialize in doing certain things) the connections engaged less frequently are actively pruned and thus weakened, while those in the circuits used more often get reinforced.[6] In short, one's every activity and experience literally change one's brain. We experience the cumulative results of this process as the relative ease or difficulty we have performing certain activities at a certain age (e.g., learning to speak a new language, read, or ride a bike). If the neural networks in relevant areas of the brain have atrophied through disuse, it takes more effort to use and re-forge them. It's not impossible, just harder, in the way that distance running, learning to play a new instrument, or writing with your non-dominant hand is challenging, but can be mastered through training and practice.

Most important for our purposes here is the fact that these neural processes help explain a person's lifelong capacity for change. Children exhibit the highest levels of malleability, but because our brains never stop learning—continuing to wire and re-wire themselves, forming, pruning, and reconfiguring synaptic structures, connections between neurons, and entire neural networks in response to our environments, circumstances, and especially interactions and exchanges with others—this plasticity never completely goes away.

All of this has led to an understanding of the mind, and the self it generates, as a self-organizing process which emerges from and regulates the flow of energy and information within the body and, crucially, within our relationships. Far from being a static self-existent entity, the self is an open, active, dynamic phenomenon, inextricable from one's interrelationships with the world and other people.

No Paradox Here

Contrary to outmoded ideas about the self as complete, closed, and standing in opposition to society, available evidence allows us now to understand the individual self as *inherently social*. And because the conditions that make up a self are arising and re-arising each moment in ways that allow for continuity *and* freshness, we naturally exhibit qualities of stability *and* change. A person is socially conditioned *and* at the same time unique, a coherent whole *and* an unfolding process. To one habituated to either/or thinking, these statements may seem paradoxical, but with a view capable of seeing wholes simultaneously as parts within complex systems, there is no such problem.

With language biased toward dualism, however, this view can be difficult to sustain. To do so, we need a concept capable of expressing the natural simultaneity of one's individuality and sociality, a term which reflects the understanding that one's character, however unique, is necessarily "a network product formed in a continuous interplay of relationships to other people" (Elias 1991:26). We need a way to communicate that the individual form of a person is a society-specific form. Happily, we already have such a concept in *habitus*.

A Helpful Alternative: Habitus[7]

Like figuration, the term habitus is probably unfamiliar to most, despite the fact that it's been around for millennia. It can be found as far back as Aristotle's writings, is lightly sprinkled throughout nineteenth- and twentieth-century European philosophy and sociology, and was made famous (at least in some academic circles) in the late twentieth century by French sociologist Pierre Bourdieu, who managed, at the same time, to make the concept somewhat infamous. The controversies generated by his use of habitus are significant enough to require mention here, after which we'll get on with exploring its basic meaning as Bourdieu, Elias, myself, and others have understood and used it. In brief, habitus represents, at a high level of synthesis, the system of schemas that constitute one's socially learned way of being in the world.

A Note on Terminology

Latin in origin, habitus comes from *habēre*, "to have, carry, wear." Sociological use of the word refers to having, carrying, and wearing the system of learned schemas which organize thought, perception, and behavior; orient activity, and operate at conscious and non-conscious levels. The suffix *-tus* transforms this verb into an action noun. The result is a term which signifies an active state of being, a subject engaging in the processes of co-creating, embodying, and expressing the complex system of self, which is relational, open, dynamic, and self-organizing—emerging from the flow of energy and information within the body and our relationships with others *while also* recursively acting on it to affect that flow of energy and information within bodies and inter-relationally.

In contrast with the static thingness conventional terms imply, Elias and Bourdieu both use habitus to denote the relational process of embodied social learning. Bourdieu, however, wielded the term far more self-consciously as a weapon against prevalent but inadequate explanations of social action via either determinism or voluntarism. He saw habitus as a means for appreciating and capturing "the principle of the continuity and regularity which objectivism discerns in the social world without being able to give them a rational basis," and at the same time "the principle of the transformations and regulated revolutions" which neither extrinsic determinants nor internal voluntarism can account for (Bourdieu 1977:82).

Bourdieu's efforts, though influential, were not altogether successful. Habitus came under harsh criticism, both for being overly deterministic and for implying excessive choice and thus explaining nothing. In part, misunderstandings were abetted by Bourdieu's prose. Within the existing linguistic and conceptual context, he had to resort to dualistic language to explain a nondual concept. His description of habitus, for example, as the "*dialectic of the internalization of externality and the externalization of internality*" (1977:72) did not make it easy to transcend the "usual antinomies" in order to be able to contain the seemingly paradoxical nature of habitus as "infinite yet strictly limited generative capacity" (1990:55).

As with the sphinx's riddle, the difficulties in grasping the nondual nature of humans as unique socially conditioned beings—shaped by and shapers of society, free to make choices within constraints, stable yet changing—derive from a failure to recognize the basic foundations of our humanness. Attending to human biology, which Elias does explicitly, provides the supplement needed to fully appreciate the habitus concept and shift our view of self, as Bourdieu advocated, "from the *opus operatum* [the work wrought] to the *modus operandi* [a process of doing something, the practice]" (1977:72).

Grounded in this knowledge, we can come to appreciate habitus as a natural process and understand what it means to describe it as:

> durable, transposable dispositions, structured structures predisposed to function as structuring structures" and "an infinite capacity for generating products—thoughts, perceptions, expressions and actions—whose limits are set by the historically and socially situated conditions of its production. (Bourdieu 1990:53 and 55)

Seeing habitus as a concept which represents a synthesis of social, biological, and physical phenomena, we can employ it, as Elias did, without having to make a big fuss about it, using it to refer to the "second nature" we acquire growing up within a certain set of conditions. With those core sociological premises as a starting point, we can understand "second nature" to be more than just an idiom referring to what feels natural and can begin to appreciate the literal meaning of this expression.

Second Nature

Thinking in terms of levels of phenomena, we can consider our biophysical bodies—with their inheritance of genes, structures, adaptive traits, and universal human needs—as the base layer of who we are, our "first nature." Through and overlaying this, we necessarily accumulate a "layer" of socially learned attitudes, perceptions, preferences, tendencies, abilities, and behaviors in ways that come to feel natural, normal, and objectively *the way it is*. Together, these make up our second nature, the socially derived qualities that are dependent on, emergent from, and expressed through the biophysical conditions that we are and exist within.

Spoken language serves as a good example. Talking is a uniquely human ability, one made possible by our shared evolutionary equipment, including a longer throat, smaller mouth, more flexible tongue and lips, and control over our breath—all of which make shaping sounds easier for us than for other species. It is only through this biophysical infrastructure that we can speak languages, systems of sounds, and meanings through which we communicate. For me, speaking English feels natural, as do the colloquialisms, facial expressions, gestures, and occasional sarcastic tones that accompany my speech. Remarkably, native speakers of other languages, with their respective flavorings, feel the same way.

This sense of naturalness is likewise the case with most other "normal" cultural activities (e.g., wearing jeans or a sari or an agbada; getting around by horse, bike, car, train, or our own two feet; having morning coffee or afternoon tea; a daily shower, a river rinse, or a bucket bath; fall hunting, underwater spearfishing, or drive-through burgers). The list of possible attitudes, perceptions, preferences, tendencies, abilities, and behaviors that can potentially contribute to our second nature is endless. The important point is that none of them is given at birth. They are learned, rather, through implicit and explicit means. Therefore, the second nature a person develops and expresses is *always* a society- (or societies-) specific form.

There is an obvious sense in which we can say that habitus formation necessarily precedes its expression. Past an initial point of human development, however, the expression of habitus is always also contributing to the ongoing process of habitus formation (reinforcing it, changing it, or both). So, for the most part, we can think of habitus expression and formation as simultaneous and interweaving processes. But because logic dictates an initial order and because of a basic distinction between the apparent *effects* of habitus formation and expression, they occupy their own positions in the model and are discussed separately below.

Habitus Formation

We come into the world ready to learn and, apart from a few universal tendencies and limitations, the range of what we can learn to think, say, like, and do is vast. From day one we participate in an endless stream of experiences and learning, from both other people and the world around us. At the level of our "first nature," this learning involves the activation of neurons, the configuration of neural networks which will ultimately support the repetition of certain activities, and numerous other processes that together culminate in habitus formation. Thanks to the sciences that study those mechanisms and processes, we can appreciate what underlies the abilities, perceptions, preferences, habits, and practices that together make up an overall way of being in the world. Importantly, though, habitus is not reducible to those.

The content available for learning exists largely in what we call culture (the collective accumulation of shared values, beliefs, symbols, norms, and material artifacts), which already exists and is manifesting in the bodies, minds, relational patterns, and stuff surrounding a person when they enter the world at birth. And, because we can only learn from those to whom we have access (whether direct or indirect) through the existing relational infrastructure, figurational structures and their qualities (i.e., size, degrees of differentiation and integration, power ratios, rates of change) have a tremendous influence on the content learned and the nature of the learning. This is why figurations and habitus are two sides of the same coin in the sociological sciences.

As "an open system of dispositions that is constantly subjected to experiences, and therefore constantly affected by them in a way that either reinforces or modifies its structures" (Bourdieu and Wacquant 1992:133), habitus is constantly in formation. Over time, the schema and embodied knowledge and tendencies acquired through social learning come to feel completely natural. Acknowledging the biophysical processes at work, we no longer have to re-hash pointless debates about nature versus nurture or try to confirm the relative importance of either. Instead, we can proceed with confidence in the knowledge that the need for the effects of nurture are decreed *by nature*. The second natures we acquire wholly depend on and emerge from our first nature. At the same time, what we learn is constantly effecting changes "down" at the biophysical level in our brains and bodies and "up" in the world at large.

Habitus Expression

The formation and expression of habitus are inherently interdependent processes—not only with each other, but also with myriad other conditions. Unlike some have misunderstood the concept of habitus to imply, there is no perfect correlation between the content learned and the way it gets expressed. These processes occur in conjunction with a unique combination of genetics, experiences, and context-specific opportunities and are therefore perpetually pregnant with possibility. So, rather than saying habitus "determines" a person's activities (e.g., thoughts, perceptions, practices, tastes, judgments), it's more accurate to say that it orients, and in some ways, engenders them. The overall result is a distinct expression of a particular sort of society-specific habitus.

Some of the activities habitus orients are more easily observable than others. One can see, for instance, characteristics of clothing, speech, bodily comportment, and other outwardly visible manifestations of a particular kind of second nature fairly directly, whereas activities of perception, thought, and judgment require some additional steps to detect. Overall, though, it is through its expression that habitus is empirically observable. At the same time, we cannot understand it without attending to the conditions which set the limits and opportunities of its formation. In contrast with conventional social science models, which seek to establish correlations between particular "internal" and "external" variables, the concept of habitus and the proposed framework naturally direct our attention to the biophysical and social circumstances underlying, conditioning, and orienting people's activities in particular ways.

Because figuration and habitus refer to two different but inseparable levels of the human social realm, we need to constantly consider both in order to understand either. When we do, we gain invaluable insights about habitus and its radical variation across place and time. One is that, observing changes in the size and complexity of the figurational structures we're embedded in, we can understand why the habitus we form while developing within those contexts has grown more complex—with multiple layers of schemas deriving from multiple sets of interrelated figurations, at varying degrees of generality. I, for example, exhibit perceptions, thought patterns, and actions respectively attributable to being: American, white, midwestern, female, from a working-class family in a rural area, a child of the 1980s, a college graduate, an academic, a permaculturist, and lots more, the expressions of which are activated and operate with varying frequency and strength according to context.

In stark contrast with past figurational conditions which involved, to varying degrees, simpler and more localized patterns of bonds of functional interdependence, most today are embedded in multiple, complex, far-reaching, highly differentiated figurational webs and linked to countless others through, for instance, schools, friends, workplaces, organizations, subcultures, ethnic groups, teams, corporations, clubs, governments, and so much more beyond kin and clan. Those connections span multiple levels of integration (e.g., local, county, state, regional, national, and international) and occur in diverse forms (e.g., face to face, print, audio and video recordings, television, online). It is a tremendous advantage, therefore, that the universally applicable principles and concepts in the socio-environmental synthesis framework enable us to study such drastically differing social conditions with a single tool.

For Example

When it comes to humans and eating, there are some universal needs and innate preferences and aversions that govern our relationship to food. Those are part of our first nature. On top of that, we acquire certain ways of perceiving, thinking, and acting which develop within particular figurational conditions to become our food-related second nature, or habitus, which can differ dramatically with that of others. Again, we'll consider highly contrasting figurational conditions to illustrate the point.

Recall the food-relevant patterns of interdependence described in Chapter 6. In modern industrialized societies, habitus forms within a complex and sprawling pattern of numerous interlinked chains of functions upon which our eating depends. This is radically different from the habitus developing within the simpler figurational conditions of a traditional foraging society. In this latter context, virtually everybody would perceive the leaves, nuts, fruits, seeds, tubers, flowers, roots, vegetables, fungi, and animals that constitute their diet as food and would have some knowledge about how and when to harvest, consume, process, and store them. As a matter of course, most would acquire the skills involved in some or all of the relevant steps, as well as the tastes and preferences available through and limited by these foods. People growing up in these societies would also comprehend and share in the symbols, ideas, and meanings that have come to be associated with particular foods and their production and consumption, and their behaviors would be guided by the norms associated with all that.

The ability to conjure up even a general picture of the sort of food-related second nature all this would engender, allows us to appreciate how very different it would be from the food-related habitus of people today, even accounting for the tremendous dietary diversity within and between modern post-industrial nations. For one thing, typical levels of ecological illiteracy today coupled with the wildly colorful packaging and products that fill our grocery and convenience store shelves translate into worlds of difference between what we and our ancestors would even recognize as food. Then there are the obvious differences in common knowledge, especially in terms of what people do (or can) know about their food: how it grows, where it comes from and in what season, who's involved in the process, how it gets to us, and more. As a final example, consider contrasting norms in these two populations. Grocery shopping, drive-throughs, individually wrapped servings, nearly unlimited options of ingredients and cuisines, for example, feel perfectly natural to us, whereas for people in more traditional societies communal labor, food sharing, and keen awareness of limitations (of seasons, geography, and sustainable harvesting rates), for instance, are integral to their sense of normal.

Other Riddles Solved, Insights Gained

Grounded in biophysical reality, the concept of habitus (in conjunction with figurations) fosters a more accurate and reliable understanding of the processes by which we become who we are, enabling us to finally grasp the reality of the socially conditioned yet unique self. In this way, it finally resolves the so-called mystery of how individual and society are linked and reveals "individualization" and "socialization" to be the same processes viewed from different angles. This basic conceptual advancement has additional positive implications for thinking about, teaching, and studying human social life.

One is that, in supporting our ability to recognize the conditioned nature of people as a scientific fact rather than an opinion that can be legitimately disagreed with, the habitus concept also undermines commonsense assumptions about the ascribed traits of individuals and groups. Even if we know better, the absence of language to explicitly acknowledge the processes that shape people and social life lends itself to conclusions about qualities being inherent to a particular ethnicity, sex, age, nationality, class, or other designation of interest. These kinds of ideas, and the habits of thought that generate them, remain central to much common sense thinking. A social science grounded in biophysical knowledge and equipped with nondual concepts would be a valuable aid for overcoming them.

Second, related to the above statuses, which conventional social science treats as universally meaningful "variables" among which statistical correlations are used to draw broader conclusions, the proposed framework and alternative concepts show that a given status or trait is not itself inherently significant. It matters only inasmuch as bonds of functional interdependence differ *along those lines*. Alternatively stated, statuses and traits only become salient aspects of habitus if the figurational patterns one develops within vary significantly according to those designations.

Finally, the concept of habitus sheds important light on the most crucial question in socio-environmental studies: How can we alter the impacts of human activities which today are unsustainable and, left unchecked, are likely to take humanity into some dangerous territory? Despite decades of social science efforts to model the role of humans in environmental impacts, "no definitive answers have been found" (Kollmuss and Agyeman 2002:240). Within the socio-environmental synthesis framework, it becomes clear that socio-environmental impacts are the result of habitus expression, which itself is the dynamic product of a way of being forming within figurational and biophysical conditions. Therefore, its development and the impacts of its expression can be more deliberately directed through changes made in those conditions.

We know that how we think about a problem largely determines our sense of what to do about it. Seeing bacteria and viruses, for example, as villainous invaders to be eradicated as swiftly as possible led to the "scorched earth" approach in medicine, the dangerous consequences of which are now becoming apparent. As medical professionals come to appreciate the dynamic relationship between a person and his or her microbiome, their approach to solving problems of illness and promoting wellness is beginning to show signs of change. Similarly, as neuroscientists, psychologists, and therapists of all sorts come to appreciate the brain and mind as complex open changeable systems, their research, diagnoses, prescribed therapies, and prognoses begin to look different. This same logic applies to our efforts to understand "environmental problems."

As long as we see humans as separate from an environment which is external to them, it's impossible to get an accurate read on what's going on, to grasp the full meaning of the laws of cause and effect, action and reaction. As a result, this view severely limits the scope of our responses to troubling situations, as we'll see in the next chapter. Only with a more realistic picture of inherently social beings operating in ways that are both *conditioned by* and *conditioning* biophysical and figurational contexts can we be equipped to really understand the socio-environmental impacts of our activities on other people and the world and, ultimately, to be able to direct and alter them in the ways desired.

Notes

1. See Elias (1991:153–237), "Changes in the We-I Balance," for an elaborated discussion of these historical processes.
2. Henry James, *Portrait of a Lady*. 1908 [1881]:287. Charles Scribner.
3. Madeleine L'Engle, *A Circle of Quiet*. 1972:32. New York: Farrar, Straus, and Giroux.
4. The Center on the Developing Child at Harvard University shows severe neglect to be at least as great a threat to health and development as physical abuse. https://developingchild.harvard.edu/resources/8-things-remember-child-development/.
5. Older estimates put that number at 700–1000 connections per second, but as of 2017, new research from the Center for the Developing Child at Harvard University confirms it to be over a million.
6. At the time of this writing, new research into microglial cells—specialized immune cells that help to maintain brain health—is helping neuroscientists understand more about one possible mechanism of this pruning (Weinhard et al. 2018).
7. The vowels in habitus are pronounced like those in platypus. I'm no Latin expert, but based on what I've been able to find, this term fits in the nominative case, fourth declension. As such, its plural form would be denoted with a "ū" (as in *habitūs*) and pronounced with a long-u sound. At some point, it may be decided that this distinction is preferable. For now, in the interest of simplicity, I use the same word, *habitus*, to denote singular and plural, as with words like fish and deer.

References

Biddle, Craig. 2014. "Individualism vs. Collectivism: Our Future, Our Choice." *The Objective Standard*, February 2. https://www.theobjectivestandard.com/issues/2012-spring/individualism-collectivism/?add-to-cart=10556.

Bourdieu, Pierre. 1990 [1980]. *The Logic of Practice*, translated by Richard Nice. Cambridge: Polity Press.

Bourdieu, Pierre. 1977 [1972]. *Outline of a Theory of Practice*, translated by Richard Nice. Cambridge: Cambridge University Press.

Bourdieu, Pierre, and Loïc J. D. Wacquant. 1992. *An Invitation to Reflexive Sociology*. Chicago: University of Chicago Press.

DeSalle, Rob, and Susan L. Perkins. 2015. *Welcome to the Microbiome*. New Haven and London: Yale University Press.
Edelman, Gerald, and Giulio Tononi. 2000. *A Universe of Consciousness: How Matter Becomes Imagination*. New York, NY: Basic Books.
Elias, Norbert. 1991. *The Society of Individuals*. New York and London: Continuum.
Gopnik, Adam. 2011. "All Alike." *The New Yorker*, January 3. https://www.newyorker.com/magazine/2011/01/03/all-alike.
Kollmuss, Anja, and Julian Agyeman, 2002. "Mind the Gap: Why Do People Act Environmentally and What Are the Barriers to Pro-Environmental Behavior?" *Environmental Education Research* 8(3): 239–260.
Markus, Hazel Rose, and Shinobu Kitayama. 1991. "Culture and the Self. Implications for Cognition, Emotion, and Motivation." *Psychological Review* 98(2): 224–253.
Montgomery, David R., and Anne Bikle. 2016. *The Hidden Half of Nature: The Microbial Roots of Life and Health*. New York and London: W. W. Norton.
Peikoff, Leonard. 2018. "Ayn Rand's Ideas, An Overview." https://www.aynrand.org/ideas/overview.
Reid, Ann, and Shannon Greene. 2013. "Human Microbiome: FAQ, Report From the American Academy of Microbiology." Washington, DC: American Academy of Microbiology.
Siegel, Dan. 2012. *The Developing Mind*. New York, NY: The Guilford Press.
Specter, Michael. 2012. "Germs Are Us." *The New Yorker*, October 22. https://www.newyorker.com/magazine/2012/10/22/germs-are-us.
Triandis, Harry. C. 1995. *Individualism and Collectivism*. New York, NY: Westview Press.
Vignoles, Vivian L., and Ellinor Owe, M. Becker, P. B. Smith, et al. 2016. "Beyond the 'East–West' Dichotomy: Global Variation in Cultural Models of Selfhood." *Journal of Experimental Psychology: General* 145(8):966–1000.
Weinhard, Laetitia, Giulia di Bartolomei, Giulia Bolasco, Pedro Machado, et al. 2018. "Microglia Remodel Synapses by Presynaptic Trogocytosis and Spine Head Filopodia Induction." *Nature Communications* 9(1228):March 26. https://www.nature.com/articles/s41467-018-03566-5.

CHAPTER 8

Actions and Reactions

"No Impact Man," Colin Beavan made a big splash in 2009 with his venture in zero impact living. Characterizing himself as a guilty liberal attempting to save the planet, he swore off plastic, fossil fuels, electricity, packaging, and more and took his wife and child along for the (bicycle) ride. Following his year-long quest were a book, a documentary, a blog, and a busy schedule of lecturing around the country. Though perhaps more sensational than most, Beavan's work was just one of numerous projects in the early decades of the twenty-first century in which people attempted to eat local, go carbon free, achieve zero waste, or live differently in some other specific way to minimize their ecological impacts.

Meanwhile, the fact that human activities could generate impacts of planetary proportions was becoming common public knowledge, and scientists were confronting the idea that human activities had ushered in a whole new geological epoch. A new genre of research investigating the processes which regulate the stability and resilience of the overall "Earth system" was emerging. One result of this work was the notion of planetary boundaries, thresholds marking the safe operating space within which human societies can sustain themselves. Surpassing these, they warned, will almost certainly bring about large-scale, abrupt, and possibly irreversible disruptions to ecosystems and the societies that depend on them. Their main conclusion: humanity "needs to become an active steward of

D. Kasper, *Beyond the Knowledge Crisis*,
Palgrave Studies on Norbert Elias,
https://doi.org/10.1007/978-3-030-48370-8_8

all planetary boundaries…in order to avoid risk of disastrous long-term social and environmental disruption" (Rockström et al. 2009).

More than just a wake-up call, this work highlighted the monumental mismatch between growing knowledge about what our socio-environmental situation calls for and our responses to it thus far. In the United States, for example, a fixation with individual behaviors has long-dominated efforts to respond to problems of planetary proportions. Even as the climate warmed, species loss accelerated, and Earth systems inched nearer to the brink of instability, things like canvas bags, reusable bottles, and efficient light bulbs continued to be touted as "solutions." Though positive, such actions have had little discernible effect on those problems. As Bill McKibben (2008) once explained, "it is true that if you clean the coils beneath your refrigerator it will run more efficiently, and it is also true that it won't do anything to 'preserve a planet similar to that on which civilization developed and to which life on Earth is adapted.'" So, demands environmental writer Derrick Jensen (2009), "why now, with all the world at stake, do so many people retreat into these entirely personal 'solutions'?"

Although critical of the stunt-like nature of Beavan's project in her review of the book, *No Impact Man*, environmental journalist Elizabeth Kolbert (2009) gets it and is somewhat sympathetic. Unless you're the Intergovernmental Panel on Climate Change (IPCC), Al Gore, or the Pope, she says, acknowledging their clout, you'd better get a gimmick. Seeing these stunts as different responses to the same major problem—that, "owing to a combination of factors…humanity is in the process of bringing about an ecological catastrophe of unparalleled scope and significance"—Kolbert understands the difficulties of trying to combat the problem in a meaningful way. Adding to the challenge is the fact that there are some seriously powerful interests deeply vested in the status quo.

In addition to all that, this chapter argues, there are even deeper reasons for the failure to calibrate our actions with what the situation demands. In particular, how we think about and experience the socio-environmental impacts of human activities profoundly shape our sense of what to do about them. We moderns have two strikes against us in this regard. Faulty mental models depicting humans and the environment as separate have made it more difficult to see the connections between what we do and the impacts of that on the world. In the context of increasingly larger, complex, and invisible webs of interdependence, the consequences of our actions are ever more difficult to see. Without clear feedback between cause and effect, action and reaction, it's even harder to generate reliable responses.

In what follows, we'll look at some conventional, but not totally accurate, ideas about the impacts of human activities, and how the double whammy of seeing them incorrectly and not seeing them at all lends itself to an extraordinary mismatch between what *needs* to be done and what most of us think we *can* or *should* do. Given the dangers we are currently courting—revealed by scientific pathway models which make for a collective and real life choose-your-own-adventure story—we know that what needs to be done, at this point, requires radical and immediate changes in economic, governance, and other fundamental social systems. Meaningful action in that direction requires broader awareness and appreciation of that basic fact.

Unfortunately, the absence of an integrated framework for understanding long-term socio-environmental processes created a vacuum that other unhelpful narratives have filled. This book seeks to address that problem by providing a theoretical framework that fosters a bigger picture, longer-term, less emotionally-involved, and more reality-congruent view of socio-environmental processes. This chapter demonstrates how a more realistic view—encompassing the relationships between physical, biological, and social phenomena in a way that precludes the otherwise constant need to re-unite them—supports an upgraded understanding of our socio-environmental impacts. With this, we are better equipped to determine effective ways to lessen our negative impacts on other people and the world and to explore the rarely asked, but perhaps more important, question: What kinds of impacts do we *want* to have?

Conventional Thinking About Impacts

As scientific evidence about the far-reaching effects of human activities on the environment was translated into public conversation, a few ideas became especially salient. One is the notion of environmental impacts itself and an awareness that what we do has ramifications for the non-human world beyond ourselves. Another involves the recognition that these impacts can be problematic. Third, and related to this, is the assumption that the ultimate goal should be to eliminate our environmental impacts. In the context of an emerging environmental public consciousness, these ideas initially represented progress. However, developing within and being expressed through conventional language and habitual thought patterns, they eventually hardened into commonsense assumptions that reflected qualities of linearity, dualism, and separateness in our view of the impacts of human activities.

"Environmental" Impacts

Turn to any environmental sciences or studies textbook and you'll find a long list of commonly featured "environmental issues" caused by human activities: air and water pollution, acid rain, ozone thinning, invasive species, deforestation, soil erosion, desertification, biodiversity loss, and more, along with a host of additional problems accompanying a changing climate. The causes and consequences vary, but one thing these impacts all have in common is that they tend to be viewed as environmental. We speak assuredly about "the environment" upon which humans inflict harm and can likewise be "friendly to," "help," and try to "save."

It's true. Human activities do have these kinds of environmental impacts. The language that expresses that awareness can be interpreted as representing an expanded sense of responsibility and thus a positive development in ethical and social thought. On the other hand, this environment-centric language bolsters the illusion of separateness, reflecting and reinforcing a very limited sense of the consequences of our impacts and who is on the receiving end. Zooming out a bit more to see where their cascades of consequences lead, we find that the impacts of our activities don't stop at some imagined border marked "the environment."

The impacts of the smoke, fumes, particulates, toxic chemicals, radioactivity, and the other substances humans produce or release, for example, are never purely "environmental," but are taken up by the bodies of beings who are always taking in, metabolizing, and *becoming* the stuff of the world. Through physical, biological, and social channels, what humans put into the world can impact human health, quality of life, and societal dynamics for years, lifetimes, or even generations to come—as can what we take from it, as ecological footprint data illustrate. This measure of how much biologically productive land and sea area we use annually in procuring resources and producing waste indicates that humanity is living beyond its means, and has been since about 1970. But not all footprints are created equal. National-level footprint data show that if everyone

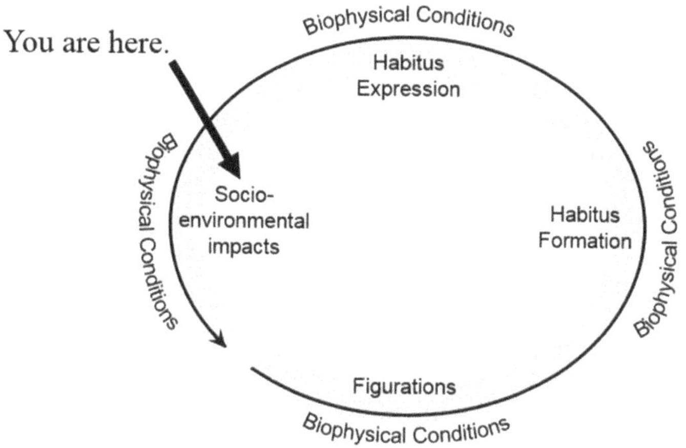

Fig. 8.1 Socio-environmental impacts. Orienting ourselves in the model, we are reminded that socio-environmental impacts emerge from human activities, which are largely the expression of the habitus that form, develop, and change within figurational and biophysical conditions

lived the average American lifestyle, for example, humanity would require nearly five earths—an impossibility that makes its unsustainability plain.

In any case, the cumulative effects of our ecological footprints go well beyond "the environment." They directly affect real people, now and in the future, saddling some with disproportionate quantities of waste and its long-term consequences and with vastly diminished resources and ecosystem services. It is therefore more accurate to think of these impacts as *socio-environmental*, hence the designation in Fig. 8.1. It's also important to note that, although physical laws make impacts inevitable, they are not inescapably negative, as conventional language implies.

Impact = Problem

The fact that climate change, mass extinction, food insecurity, soil degradation, water pollution, and other major crises are the cumulative results of human behaviors leads some to conclude that human activities—and by association humans themselves—are not compatible with sustaining healthy ecosystems. Some have explicitly stated, while others quietly harbor a suspicion, that the world would be better off without us. At the very least, the goal of minimizing human impacts is widely accepted. Despite the distinctively negative connotation that the phrase "environmental impact" presently has for most people, the truth is that the consequences of our actions need not be all bad.

To the contrary, the certainty that what we do will have an impact on other people and the world—exercised with the benefit of a more accurate understanding of ourselves and our place in the world—reveals the possibility of more consciously guiding human impacts in ways that are actually beneficial to humans and other species. When we flip our thinking in this way and begin to imagine the positive socio-environmental impacts we *could* have, "then impact is a great thing, a footprint is something you want to leave," says farmer and system designer Ben Falk (2017). Taking this idea further, permaculturist Geoff Lawton (2019) argues that, by aligning our actions with natural systems, we can "become the most beneficial element on Earth, rather than the most damaging." It's just a matter of design.

Engaging in this kind of design requires us to employ our unique human powers of inquiry, investigation, and imagination alongside hard-won knowledge in service to two important questions: *What kinds of impacts are we having? And what kinds of impacts do we want to have?* We've gotten pretty good at answering the first question. And, in response to the harms identified, a general aspiration to mitigate and ultimately eliminate them has emerged. In many ways, this is a natural and healthy human impulse. Being overly fixated on this goal, however, keeps us from considering the positive impacts our activities *could* have. Even worse, it reinforces a third misguided assumption: the possibility of a zero impact life.

Have No Impact

When the dominant cultural trope of "going green" means reducing one's impact as much as possible, taking it to its logical conclusion means eliminating them altogether. Remember, Beavan "didn't just want to have no carbon impact. [He] wanted to have no environmental impact" (2009:14). Considering the cosmic statute of cause and effect, this is just plain impossible. When the ideal aspired to is unattainable, failure is inevitable. And failure is not very encouraging. Like a dieter who gives up after a day of overindulging, aspiring greens, upon discovering that they can't do it perfectly, are liable to give up on their efforts toward a no impact life. The whole thing just seems too hard and, as long as they see it as an all or nothing game, they would be right. Not only has this wrongheaded idea caused trouble for well-intentioned people who long to live environmentally and socially responsible lives, it also helps perpetuate inaccurate notions of how the world works.

The laws of physics ensure that all causes have effects, and all actions have reactions. As organisms existing within biophysical conditions, our activities necessarily have impacts. Paleontologist and earth scientist Tim Flannery reminds us of this in *The Weather Makers*, where he poetically explains that everything we do, from the mere act of breathing on up, is registered in the biosphere we share with untold numbers of beings.

> The air you just exhaled has already spread far and wide. The CO_2 from a breath last week may now be feeding a plant on a distant continent, or plankton in a frozen sea. In a matter of months, all of the CO_2 you just exhaled will have dispersed around the planet. Because of its dynamism, the atmosphere is on intimate terms with every aspect of our earth, from the mantle upward. No volcano belches, no ocean churns—indeed, no creature breathes—without the great aerial ocean registering it. (2005:22)

The effect of one breath is miniscule. But when we add up the "breathing" of over a billion passenger cars, millions of trucks, and about 100,000 flights per day in just one sector of human activities, we can begin to appreciate the massive impacts those activities have altogether.

Early on in climate change discourse, some suggested that humanity was too small to make a difference on so vast a planet. They were wrong, of course, but should anyone doubt that our species could have a discernible impact, they need only consider the fact that organisms trillions of times smaller than ourselves were capable of transforming Earth's atmosphere and the trajectory of life on this planet. Anaerobic organisms were once the dominant form of life on Earth, but thanks to an evolutionary twist, organisms capable of making energy from sunlight came onto the scene. The by-product of this novel self-sufficiency was oxygen and, over time, cyanobacteria raised oxygen levels to a point that sent their anaerobic cousins packing, setting the stage for the evolution of life as we know it. This example reminds us that living beings, no matter their size, have impacts which, in numbers great enough and over time periods long enough, can alter the fate of an entire planet. Although it took cyanobacteria billions of years to transform the Earth, humans are achieving this feat in a tiny fraction of the time. The more important difference, though, is that *we know it.*

As far as we can tell, *Homo sapiens* is the only species capable of that kind of knowing and of exercising self-reflexive awareness of the transformative impacts of our activities and their long-term consequences for other beings. Alas, even as mounting scientific evidence adds ever greater detail to our understanding of those impacts, conventional thought patterns keep us ill-equipped to take the kinds of actions the situation calls for. Within the socio-environmental synthesis framework, however, we can recognize and correct key errors in our thinking in support of the ultimate goal of altering the socio-environmental impacts of human activities in the ways deemed necessary at this crucial juncture in human history. In order to appreciate what's necessary, we need to first establish a shared understanding of our situation by taking a look at the overall state of the system (and its subsystems) on which human societies depend.

Earth Systems in a New Epoch

What scientists call the "Earth system"—the planet's interacting physical, chemical, and biological processes—has undergone changes over long periods of time due to changing circumstances and natural fluctuations in the systems that comprise it. Relative to human history, though, its patterns have appeared to humans to be pretty much fixed. Using the language of Chapter 5, we can think of the Earth system and its overall functioning in the Holocene epoch as a set of relatively fixed biophysical conditions. It is this stability, we now recognize, that has made modern human civilizations possible.

Within the current Quaternary period (the last 2.6 million years, characterized by the cyclic growth and decay of continental ice sheets associated with fluctuations in the Earth's tilt and orbit around the sun), the Holocene epoch began around 11,700 years ago with the end of the last ice age. Despite some temperature variation, the Holocene is characterized by a fairly stable climate, warming that expanded habitable zones northward and corresponding shifts in the distribution of plants and animals. It is generally agreed that warmer temperatures and relative climatic stability are what allowed humans to reliably engage in agriculture and to establish, and rapidly expand, permanent settlements. These activities, of course, had impacts.

Robust stable systems can withstand a certain amount of disturbance and change and still function in their usual way, or at least quickly return to it. The human body's thermoregulation system, for example, keeps us at a pleasant average of about 98.6°F,[1] without much conscious effort on our parts. In situations of high heat or cold, negative feedback loops kick in, contracting or expanding our blood vessels, inducing sweating or shivering, and bringing us back into normal range. But systems have tipping points, thresholds beyond which they cannot return to normal functioning even when the disturbance has been removed. Extreme conditions like a plunge into icy waters or a very high fever can push our body temperature out of normal functional range causing the body to suffer injuries to tissues and organs, brain damage, or death. Though larger and more complex than a single organism, the Earth system works in much the same way. There are limits to the amount of change and disturbance it and its subsystems can withstand and still continue to function "normally."

The Anthropocene

As our numbers and the scale of our activities expanded, humanity's impacts on the planet's systems became increasingly potent, eventually pushing the overall Earth system into a new state, one in which human activities are a significant driver of planetary changes, some of which have left their mark on the geological record. Many scientists came to think of this novel situation as significant enough to warrant its own designation. This distinct epoch has come to be known as the Anthropocene.

The idea that humans are a force of planetary change is no longer controversial. What scientists cannot agree on, however, is exactly *when* the shift took place. Some have identified the emergence of wet agriculture (e.g., rice paddies, taro crops, weir construction) and the cutting and burning of forests as the main factors that tipped the balance in Earth's systems, leading them to place the start of the Anthropocene at about 8000 years ago (Flannery 2005). Others argue that, while the imprint of early human activities can be seen at a regional scale, there is no clear evidence that humans have impacted the functioning of the Earth system at the global scale until more recently, causing them to place the threshold of the Anthropocene at the rise of the industrial revolution (Rockström et al. 2009; Steffen et al. 2007). Still others say that, although the advent of fossil fuel use is of unquestioned significance, "the evidence of large-scale shifts in Earth System functioning prior to 1950 is weak" (Steffen, Broadgate, and Deutsch et al. 2015:13). They pinpoint July 16, 1945, as the beginning of our new epoch. As the date that the first atomic bomb was detonated in New Mexico (emitting radioactive isotopes into the atmosphere from which they entered the sedimentary record), it provides a unique geological indicator and marks the start of a distinct phase in which the strength and reach of human impacts were greatly accelerated. Aptly, they named this phase "the Great Acceleration."[2]

Regardless of where we place the start time, the following fundamental facts are sobering. The Holocene is the only known state of the planet that can support the world as we know it (Rockström 2015). Without the pressures humans are presently exerting on Earth systems, the Holocene state could likely be maintained for thousands of years into the future (Berger and Loutre 2002). At present, humans are pushing the planet outside the Holocene range of variability for many key Earth system processes (Steffen, Richardson, Rockström et al. 2015). Taken together, these findings prompted an important and novel question: "What are the non-negotiable planetary preconditions that humanity needs to respect in order to avoid the risk of deleterious or even catastrophic environmental change at continental to global scales?" (Rockström et al. 2009).

Courting Dangers, Comparing Futures

Ironically, while our methodological and technological capacities to gather increasingly sophisticated information about planetary boundaries have grown, our position with respect to those boundaries has worsened. Between 2009 and 2015 (the publication year of the team's follow-up paper), atmospheric concentrations of carbon dioxide increased by almost 15 parts per million, a trend that has since continued. Data indicated that the status of four main Earth system processes had worsened; by 2015, climate change, genetic diversity, land system change, and biochemical flows were all found to have exceeded the proposed boundary of safety (see Fig. 8.2).

Whatever else we might say about the kinds of impacts we want to have, the planetary boundaries framework clearly suggests that, at the very least, one collective goal must be to return to, or remain within, the safe (i.e., functional) zones of Earth systems. Moving outside of those is likely to bring about collapse of the very systems human societies rely on. While

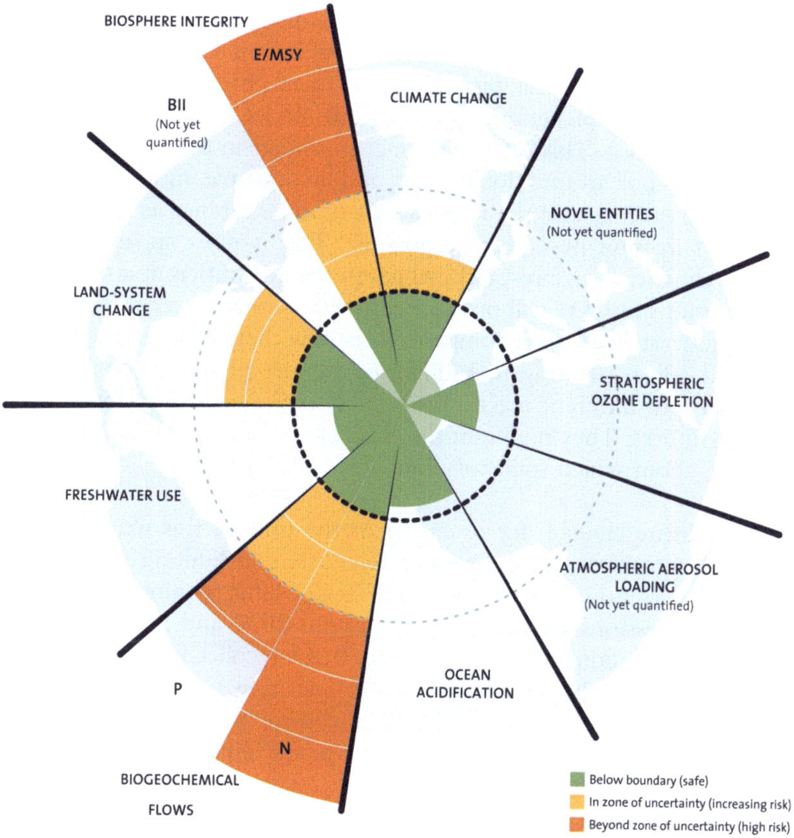

Fig. 8.2 Graphic depicting the status of eight planetary boundaries, as of 2015. The green zone is the safe operating space, the yellow represents the zone of uncertainty and increasing risk, and the red is a high-risk zone. The planetary boundary itself lies at the intersection of the green and yellow zones (*Credit* J. Lokrantz/Azote based on Steffen, Richardson, Rockström et al. 2015)

societal collapse doesn't sound good, it's important to remember that there is nothing especially novel about it. Historical evidence shows it to be more the rule than the exception. What *is* new is the magnitude of collapse we are ushering in. The potential biophysical and societal system breakdowns are now global rather than regional in scale.

Along with the scale of our predicament, the degree of knowledge we have about it is also historically unique. Never before have humans known so much about the physical, biological, and social processes relevant to the functioning of the planet and been able to model and compare various "what if" scenarios. This offers a distinct advantage in terms of being able to look back, look around, look ahead, and, in principle, more consciously shape our trajectory and the implications of our existence as a species on this planet in ways never before possible. This is most apparent in the realm of climate change research, which shows how different atmospheric concentrations of CO_2 will produce different warming scenarios. One of the greatest values of such comparisons is how they illustrate the laws of cause and effect, showing how different avenues of action (or inaction) today will spur different reactions in our system, generating dramatically different futures. They highlight the choices we have made, and those we still have, about which trajectory humanity, and everybody else, ends up on.

Take climate change, for example. At the time of this writing, there is general agreement among climate scientists that limiting warming to the 1.5°C currently deemed the "safe" upper limit requires that greenhouse gas emissions rapidly decline between 2020 and 2030, with net CO_2 emissions reduced to zero by 2050. Our collective failure to act earlier and decisively on climate change means we must now achieve even deeper emissions cuts. Had serious climate action begun in 2010, one authoritative comparison shows, the average annual emissions reductions required to achieve emission levels in 2030 consistent with a 1.5°C warming scenario would have been 3.3%, as opposed to the 7.6% per year cuts required now as a result of dragging our feet on climate (UNEP 2019).

Such bold action, analysts agree, would demand unprecedented efforts to transform our energy, economic, governance, cultural, infrastructural, and other social systems. To achieve this, "the world must embark on a World War II-level effort" (Nuccitelli 2018). Delayed action, limited international cooperation, and weak policies, on the other hand, would put the possibility of limiting global temperature rise to 1.5°C out of reach (Rogelj et al. 2018; UNEP 2019).

Given that current national *ambitions* for greenhouse gas mitigation through 2030 are consistent with global warming of about 3°C by 2100 with continuing warming afterward (IPCC 2018) and that *actual* global greenhouse gas emissions to date continue to rise, all signs indicate that we are in the process of foregoing a historically rare opportunity to apply the vast and growing stock of human knowledge to the vital work of figuring out how to integrate the design and development of human societies into the maintenance of resilient and accommodating Earth systems (Steffen, Richardson, Rockström et al. 2015). This is the sort of missed opportunity that prompted scientists from around the world to put out a "second notice" on their 1992 warning about the drastic changes needed "if vast human misery is to be avoided and our global home on this planet is not to be irretrievably mutilated" (Union of Concerned Scientists 1992). Sadly, data show that over the course of the ensuing 25 years, with the exception of stabilizing the stratospheric ozone layer,

> humanity has failed to make sufficient progress in generally solving these foreseen environmental challenges, and alarmingly, most of them are getting far worse...Soon it will be too late to shift course away from our failing trajectory, and time is running out. (Ripple et al. 2017:1026)

Scientific data are invaluable for providing us with details about the impacts of human activities on core Earth systems and for indicating *what* needs to change in order to stay within planetary boundaries of functionality for humans. Science cannot, however, tell us exactly how to go about doing it. As David Wallace-Wells puts it in *The Uninhabitable Earth*, "the question of how bad things will get is not actually a test of the science; it's a bet on human activity. How much will we do to stall disaster, and how quickly? Those are the only questions that matter" (2019:219).

So, What Have We Been Doing?

In reality, there are legions of people around the globe working hard in a multitude of ways to forestall climate disasters and address other socio-environmental problems. But without sufficient coordination, it's difficult to imagine how they might add up to the "World War II-level effort" and humanity-wide stewardship of planetary boundaries we're told is necessary to avoid disastrous and long-term disruption. This degree of coordination and collaboration is difficult to imagine given the types of action which have dominated so far, none of which has delivered the results needed.

From the Top

Since at least the 1970s, there have been efforts to mitigate humanity's socio-environmental impacts at the level of international negotiations. In 1972, the United Nations (UN) held its first major conference on international environmental issues and in 1983 established a special commission to report on them and on strategies for sustainable development. In 1992, the UN Conference on Environment and Development (aka "the Earth Summit") was held in Rio de Janeiro with the goal of reconciling economic development with environmental protections. Among its main documents was the United Nations Framework Convention on Climate Change (UNFCCC). This treaty would require nations to reduce their emissions of carbon dioxide, methane, and other "greenhouse" gases, but stopped short of setting binding reductions targets.

The Kyoto Protocol, a 1997 addition to the UNFCCC, committed its signatories by setting internationally binding emissions reduction targets. It went into force in 2005 with targets that were individually tailored to participating parties; the goal was for industrialized nations to cut greenhouse gas emissions by 5.2% below 1990 levels during a five-year commitment period (2008–2012). Although this represented a huge diplomatic accomplishment, it had some major shortcomings. First, the world's top two emitters of greenhouse gases, China and the United States, were not participating (China was exempt because of its status as a developing nation and the United States never ratified the treaty). Moreover, studies conducted in the first couple of years after it took effect showed that most participants would fail to meet their emissions targets. Finally, critics argued, even if the targets *were* met, the benefits would be insignificant because the emission reduction goals were too modest to matter over the long term.

The subsequent UNFCCC "Paris Agreement," adopted in 2016, was meant to improve upon and replace the Kyoto Protocol. The long-term goal was to keep the increase in global average temperature to below 2°C above pre-industrial levels (limiting the increase to 1.5°C if possible) and to achieve this with enhanced support to assist developing countries. As of 2019, 195 UNFCCC members had signed the agreement and 185 had become party to it (the United States, however, declared its intent to leave at the first opportunity, which it did in late 2019). Still, the agreement is neither binding nor enforceable, and even if it were, according to most analyses, it would not be enough. Corinne Le Quéré, director of the Tyndall Center for Climate Change Research, states simply, "It's not fast enough. It's not big enough," and "there's not enough action" (Dennis and Mooney 2018). While achieving consensus among nearly 200 world leaders is a genuine achievement, international diplomacy has not yet produced meaningful climate action. Interestingly, the deepest challenge to the effectiveness of these agreements, says climate scientist Kevin Anderson (2015), is not the naysayers, but a cadre of well-meaning scientists, engineers, and economists who are banking on high tech fixes. The widespread assumption that emerging technologies will play a starring role in reducing emissions without compromising growth has diminished the effectiveness of these high-level discussions.

In the Streets

As this book was taking shape, so were some new kinds of activism. The second decade of the twenty-first century saw an emerging brand of global activism that could potentially put the pressure on corporate and government bodies needed to achieve more robust action. International organizations like 350.org, Climate Action Network, Friends of the Earth, and the People's Climate Movement honed their respective strategic foci and become increasingly active *and* interactive. With the shared goal of avoiding the worst and charting a course to the best of possible futures, they staged global climate demonstrations demanding not only action, but a just transition. At present, the movement continues to grow, joining forces with youth-led climate activist groups and growing numbers of young people taking their national governments to task, and to court—suing for their failure to address climate change and its known risks.

In some ways, this movement looks like a continuation of environmental activism in 1970s America, which was instrumental in motivating the US government to create the EPA and other landmark environmental legislations. But in other ways, it looks different. It's more spread out, for one thing, with demonstrations distributed around the world instead of concentrated in one place. It's also more complex, attempting to redress a more diverse range of problems. Woven into its climate and ecological priorities are concerns about environmental justice, racism, gender equality, indigenous rights, and poverty, reflecting a growing appreciation of the *socio-environmental* nature of the impacts and problems in question. Unfortunately, another difference is the degree to which these issues have, in contrast to 1960s and 1970s United States, come to be seen as firmly partisan. The overall influence has been to further constrain potential political support and firm up relationships between polluting industries and their political allies.

This burgeoning socio-environmental movement demonstrates an appreciation of the need for immediate and aggressive high-level policy change, the achievement of which requires heightened awareness and concern among at least a critical mass of individual citizens, enough to put pressure on decision-makers at the highest levels. There's some evidence that it's having an effect. Competition among democratic presidential nominees in 2019, for instance, revealed multiple candidates with solid climate policy proposals, a first. "The thriving grassroots, youth-led movement," says *Scientific American* writer, "could push the next president to be even more visionary and move even faster toward zero and then to negative emissions" (Johnson 2019). How citizens and their political representatives ultimately respond to this brand of action, though, remains to be seen.

Up to You

Most salient among the main categories of action is the focus on individual and household behaviors as the way to "save the earth." Based on the assumption that lots of individuals making better choices will add up to broader changes in our socio-environmental impacts, this idea permeates current social thought.

When my youngest child was five, his teachers (as so many teachers across the country do each April) organized a series of Earth Day activities. One of them involved having students explore the environmental impacts of their individual actions and share ideas for how to "protect our earth." Written in their adorable fledgling penmanship, children's responses included things like "never litter," "make sure to put stuff in recycling," "turn off the water while I brush my teeth," and other familiar recommendations. These were posted on classroom doors with cutout pictures of them holding up little Earths (see Fig. 8.3).

It's an important lesson to teach children: everything we do has effects beyond ourselves; be mindful of that and act accordingly. While that's a good place to start a person off on their journey of responsible living, it's a terrible place to stop. Unfortunately, on the level of mainstream public

Fig. 8.3 Children's pledges: "I promise to protect our earth by..." (*Photo credit* Author)

discourse in the United States, this is where we have stopped and been stuck for quite some time.

When it comes to how citizens can engage in environmental action, the information coming from popular books, magazines, newspapers, and websites (even of prestigious environmental organizations) has long been dominated by lists of "easy ways you can save the planet," the easier the better. Full of tips about recycling, saving water, unplugging, changing light bulbs, and the like, they sound pretty much like what our five-year-olds are saying, and what students continue to say throughout the rest of their schooling years and into their adult lives. In essence, it seems that when it comes to understanding our socio-environmental impacts, we have failed to advance beyond the kindergarten level.

Michael Maniates has characterized this arrested development as the "individualization of responsibility," in which citizens have mostly come to see "environmental degradation as the product of individual shortcomings" (2001:32). With this as a premise, the "solutions" easily arrived at primarily involve education and enlightened consumer choice. Shored up by deep-seated individualism, this view has been cultivated by decades of targeted industry efforts focused on shifting our sense of responsibility away from firms and institutionalized practices and toward individuals (Assadourian 2013; Strand 2008; Leonard 2013). "It worked," Leonard points out. "Over the last few decades, the theme of the individual's role in wrecking the environment, and the individual's responsibility in fixing it, has only grown stronger" (2013:244).

Rousing people's awareness and sense of ethical responsibility is unquestionably a good thing, but when it's the main focus, citizens are distracted from the real drivers of their impactful activities: the policies, business practices, cultural norms, and the economic, regulatory, and physical infrastructures that presently facilitate environmentally destructive behaviors. In the absence of a comprehensive, coherent, empirically-grounded, and consistently conveyed framework for studying the dynamics between people, socio-environmental processes, and social change, the "easy ways to save the planet" narrative has risen to prominence. More to the point, green consumerism has had no discernible effect on reversing the trends in question. In fact, it may very well work *against* efforts to effect change.

For one thing, when the fate of the world seems to rest upon individuals and their daily decisions and behaviors, the pressure of having to choose between paper or plastic, cage-free or free range, and organic or local can be too much, even for those inclined to care. For the majority not so inclined, there is no decision to be made, at least not for social or environmental reasons. Factors of cost, comfort, convention, and convenience prevail. Furthermore, when environmental responsibility is portrayed as and perceived to be a personal lifestyle choice, choosing to behave in ecologically conscious ways seems like just another means to craft an identity. When choices like these are associated with certain social or political identities, then choosing *not* to behave in these ways is also an important aspect of claiming and expressing those. Finally, there's the irony of fighting problems related to excessive consumption with more consumerism. "Green" or not, most goods still represent the extraction of resources, the embedded energy in production and transport, and the disposal of the packaging and eventually the product itself. Besides, efforts to initiate significant changes toward sustainable consumption, to date, have not had much success. Despite an academic tradition of more than 25 years, one sustainability network noted,

> research on how to bring about systemic changes towards sustainable consumption and production is still rather limited…Knowledge is missing on how to bring about systemic change; and how to create conditions under which sustainable lifestyles can emerge. (KAN SSCP 2019)

The scale of change necessary to actually make a difference is daunting. As William Rees, co-creator of the ecofootprint concept, points out, achieving ecologically sustainable consumption ultimately means cutting global energy and material consumption in half over the next decade, with higher cuts required of high-income countries. Driving home the point, Rees asks readers, "will you be using 6-10% less of everything by this time next year?" (Rees 2019).

Where Do We Go from Here?

The bottom line is that, through the cumulative decisions, actions, and inaction of world leaders, we have squandered the chance humanity has had over the past few decades to gradually ease ourselves into a more sustainable position. The result has been a vastly diminished capacity to forestall disaster in any meaningful way. It is still possible to act in ways that ensure a livable civilization, but if those actions are going to matter, our responses must be equal to the task. While the above approaches (i.e., changing the rules of the game through international policy, mobilizing the demand that will motivate leaders to act, and aligning behaviors with sustainability) are important parts of an overall plan of action, their potential power is limited by conventional thinking about impacts. A fairly pervasive overemphasis on entities and events, dualistic divides, linearity and reductionism, presentism, and an overriding sense of "us versus them" has kept us running in circles without really getting anywhere.

Given the many influences on what people do, figuring out where to most effectively intervene to alter the socio-environmental impacts of human activities is no small undertaking. Along with external circumstances, prevailing rules (formal and informal), personal proclivities, universal human foibles (like ego, greed, fear, jealousy, closed-mindedness), and the equally universal imperative to cooperate all profoundly affect human actions. How these manifest, however, and which tendencies are strengthened or suppressed, is largely a matter of cultural conditioning, which itself is largely a matter of the paradigms which shape our understanding of the world—the shared ideas, unstated assumptions (unstated because everyone already "knows" them), and deep beliefs about how the world works.

Among the suite of actions we should be taking to alter our impacts, one must be to establish a more a more accurate, integrated, systemic view of reality, especially of the relationships between physical, biological, and social phenomena. This is the leverage point with which this book is mainly concerned. As systems thinker Donella Meadows famously pointed out, "people who have managed to intervene in systems at the level of paradigm have hit a leverage point that totally transforms systems" (2008:163). These collective ways of seeing and thinking about the world, she explains, are at the heart of our systems. System goals, information flows, and other system features come from them.

For a single person, paradigms can change rather quickly, in the blink of an "a ha moment," when things mentally click into place in new ways and one begins to see the world differently. Social systems are another matter. In the hierarchy of leverage points, Meadows tells us, the higher it is, the greater the resistance to change. Thankfully, she points out, *the* single highest leverage point we have at our disposal is the ability to transcend paradigms. This ability, related to Elias' concept of "detachment," is rooted in our capacity to consider a thing from a greater mental distance, freer from personal feelings and preferences, and then to retain or let go of it based on the conclusions arrived at.

As we saw in Chapter 3, Elias credits this capacity for advancements in the natural sciences, which required a loosening of our attachment to certain ego-serving but outdated ideas (e.g., the sun revolves around the Earth, all creation exists for the benefit of humans). This increasing capacity for detachment was aided by the more expansive temporal view gradually adopted in the various natural sciences. In contrast, Elias (1987) argues, this trend did not affect the social sciences in the same way. In fact, one can discern a temporal narrowing of sociologists' interests during the twentieth century, at which time it also became increasingly difficult to distinguish scientific from personal and emotional interests. In this sense, their outlook more closely resembled that of society at large, where national leaders and their populations are unable to act on established knowledge about the dangers their activities pose because their involvement in the affairs of their own state is too strong and their capacity for detachment too weak. The concepts of involvement and detachment offer a useful way to comprehend this dynamic: higher perceptions of danger breed higher levels of involvement, which in turn breed higher danger, in a reinforcing feedback loop.

Elias emphasizes the important role sociologists have to play here, especially in the creation of models depicting long-term social processes. "How far it is possible under present conditions for groups of scientific specialists to raise the standards of autonomy and adequacy in thinking about social events and to impose upon themselves, the discipline of greater detachment, only experience can show," but how else can we break the vicious cycle in which high affectivity of ideas and low ability to control dangers coming from humans to humans reinforce the problem? (Elias 1956:252). One way is to provide said specialists with tools for achieving the broader perspective and paradigm shift needed. The socio-environmental synthesis framework is one such tool.

Understanding Socio-Environmental Impacts Through the Synthesis Framework

Paradigms matter. How we perceive the world and what we believe about it shape our actions, reactions, and sense of what's possible. In the context of concern about socio-environmental impacts, they "matter for the way in which questions are framed...[and] for the types of policy intervention proposed in response to major challenges" (Shove and Walker 2014:55). We know that, despite ever higher levels of awareness, concern, and knowledge about Earth systems and the impacts of human activities, "past and current mitigation efforts have been insufficient" (UNEP 2019:39). Coming to see these impacts differently—as not just environmental but also inherently social and as both the result and basis of ongoing processes—is a crucial step in being able to change them in the ways needed.

Just as paradigms of planet Earth have shifted with the kinds of maps and images of it available, systems modelers argue that "we can change paradigms by building a model of the system, which takes us outside the system and forces us to see it as a whole" (Meadows 2008:164). Representing the hierarchy of physical, biological, and social phenomena, the socio-environmental synthesis framework depicts a general pattern of interdependent processes where the main elements—biophysical conditions, figurations, habitus formation and expression, and socio-environmental impacts—are distinguishable but never separate. It does not permit us to deny or forget that, by virtue of our very existence, we *are* and *participate in* biophysical conditions and thus that impacts are inevitable, but need not be all bad.

In the examination of socio-environmental impacts, Fig. 8.1 directs our attention backward to their origins in the collective expression of particular aspects of habitus and its formation within given figurational and biophysical conditions. It also prompts us to look forward to the potential and probable effects that particular socio-environmental impacts will have on people and non-human biophysical conditions, how those might affect subsequent figurational changes, how those changes will or could shape future habitus, and so on. Importantly, in emphasizing the role of figurations in the formation of the perceptions and practices most responsible for the impacts in question, the framework can highlight specific relational patterns as potential targets of change efforts. In this way, it generates novel insights about potential entry points, setting us up to realize existing possibilities for more intentionally guiding the direction and nature of our immediate and long-term socio-environmental impacts in ways that minimize harms and maximize desired benefits.

In contrast to the more typical approach of direct intervention, the focus here is on understanding the conditions which give rise to a given impact and then considering ways to alter them or create new ones to change impacts in the ways sought. Additionally, the more expansive time-frame it promotes supports the development of the greater capacity for detachment that the work of our time demands. Only with a deeper sense of time and our position in a much longer trajectory can we appreciate "the astonishing responsibility of legacy-leaving," nature writer Robert Macfarlane (2019) explains, prompting us to ask, as renowned medical researcher Jonas Salk (1992) did, "are we being good ancestors?" These questions are more relevant than ever in the Anthropocene, when present-day human activities will shape others' futures for a long time to come.

The intention is not to suggest that this framework, and the upgraded paradigm it supports, allows for total control or makes change easy. The tasks before us remain formidable, but we have in this framework something we have not had before: an "integrated systems approach to synthesize data from diverse fields" (Stafford 2010), theories and methods "based on the fundamental interactions between people and the biophysical environment" (Moran 2010), and a reliable theory of long-term socio-environmental processes, ideally useful for "navigating the world back into a safe operating space" (O'Brien 2015). Overall, the socio-environmental synthesis framework supports the abilities humans most need, which Salk eloquently described.

> In the realm of human consciousness the highest and most sophisticated form of self-regulation is based on our ability to see ahead. It requires a knowledge of self and the cosmos, and of self *in* the cosmos. The capacity exists in us...But it has to be evoked and developed. It is among the characteristics of the human being and of human consciousness which have been selected for and possess evolutionary value. The evolutionary need is to increase our breadth of consciousness as human beings, to expand our range of choice for the wisest alternatives. (1983:114)

Change will occur and a future that is different from the present will unfold no matter what we do. Humans will never have complete control over these processes. However, being able to understand and envision them more accurately, the subject of the next chapter, puts us in a position to direct them more intentionally, ideally enabling us to leave a better legacy.

Notes

1. This has long been the accepted average, but recent research suggests that averages today are between 0.5 and one degree lower.
2. Graphics representing these findings are featured in Chapter 1.

References

Anderson, Kevin. 2015. "The Hidden Agenda: How Veiled Techno-Utopias Shore Up the Paris Agreement." https://kevinanderson.info/blog/the-hidden-agenda-how-veiled-techno-utopias-shore-up-the-paris-agreement/.

Assadourian, Erik. 2013. "Re-engineering Cultures to Create a Sustainable Civilization." Pp. 113–125 in *Is Sustainability Still Possible?*, edited by Worldwatch Institute. Washington: Island Press.

Beavan, Colin. 2009. *No Impact Man*. New York: Farrar, Straus and Giroux.

Berger, A., and M. F. Loutre. 2002. "An Exceptionally Long Interglacial Ahead?" *Science* 297(5585):1287–1288.

Dennis, Brady, and Chris Mooney. 2018. "Countries Made Only Modest Climate-Change Promises in Paris: They're Falling Short Anyway." *Washington Post*, February 19.

Elias, Norbert. 1987 [1983]. *Involvement and Detachment*, translated by Edmund Jephcott. Oxford: Basil Blackwell.

Elias, Norbert. 1956. "Problems of Involvement and Detachment." *The British Journal of Sociology* 7(3):226–252.

Falk, Ben. 2017. In "Inhabit: A Permaculture Perspective," a documentary by Costa.

Flannery, Tim. 2005. *The Weather Makers: How Man Is Changing the Climate and What It Means for Life on Earth*. New York: Atlantic Monthly Press.

IPCC. 2018: Summary for Policymakers. In: *Global Warming of 1.5°C. An IPCC Special Report on the impacts of global warming of 1.5°C above pre-industrial levels and related global greenhouse gas emission pathways, in the context of strengthening the global response to the threat of climate change, sustainable development, and efforts to eradicate poverty* [Masson-Delmotte, V., P. Zhai, H.-O. Pörtner, D. Roberts, J. Skea, P.R. Shukla, A. Pirani, Moufouma-Okia, C. Péan, R. Pidcock, S. Connors, J.B.R. Matthews, Y. Chen, X. Zhou, M.I. Gomis, E. Lonnoy, Maycock, M. Tignor, and T. Waterfield (eds.)]. World Meteorological Organization, Geneva, Switzerland.

Jensen, Derrick. 2009 "Forget Shorter Showers: Why Personal Change Does Not Equal Political Change." *Orion Magazine*. July/August. http://www.orionmagazine.org/index.php/articles/article/4801/.

Johnson, Ayana Elizabeth. 2019. "5 Things That Went Right for Climate Action in 2019." *Scientific American*. December 30. https://blogs.scientificamerican.com/observations/5-things-that-went-right-for-climate-action-in-2019/.

KAN SSCP. 2019. "Enabling Changes Toward Sustainable Lifestyles and Resilient Livelihoods." Memo from Future Earth's Knowledge Action Network, Systems of Sustainable Consumption and Production. Email correspondence on SCORAI (Sustainable Consumption Research and Action Initiative), March 11.

Kolbert, Elizabeth. 2009. "Green Like Me." *The New Yorker*. August 31:70–74.

Lawton, Geoff. 2019. "Green Is the New Silver (Lining): Crisis, Hope and Permaculture." Video. https://www.discoverpermaculture.com/video-1-pdc-2019.

Leonard, Annie. 2013. "Moving from Individual Change to Societal Change." Pp. 244–252 in *Is Sustainability Still Possible?* Worldwatch Institute.

Macfarlane, Robert. 2019. "The Hidden Human Depths of the Underland." On Being interview with Krista Tippett, November 4.

Maniates, Michael. 2001. "Individualization: Plant a Tree, Buy a Bike, Save the World?" *Global Environmental Politics* 1(3):31–52.

McKibben, Bill. 2008. "Multiplication Saves the Day." *Orion*, November 27. https://orionmagazine.org/article/multiplication-saves-the-day/.

Meadows, Donella. 2008. *Thinking in Systems: A Primer*, edited by Diana Wright. White River Junction, VT: Chelsea Green Publishing.

Moran, Emilio. 2010. *Environmental Social Science: Human Environment Interactions and Sustainability*. Malden, MA: Wiley-Blackwell.

Nuccitelli, Dana. 2018. "There's One Key Takeaway from Last Week's IPCC Report: Cut Carbon Pollution as Much as Possible, as Fast as Possible." *The Guardian*. October 15. https://www.theguardian.com/environment/climate-consensus-97-per-cent/2018/oct/15/theres-one-key-takeaway-from-last-weeks-ipcc-report.

O'Brien, Karen. 2015. Great Transition Online Network Discussion, March 23. https://www.greattransition.org/contributor/karen-obrien.

Rees, William. 2019. Email Correspondence on SCORAI (Sustainable Consumption Research and Action Initiative), March 7.

Ripple, William, Christopher Wolf, Thomas Newsome, et al. 2017. "World Scientists' Warning to Humanity: A Second Notice." *BioScience* 67(12):1026–1028. https://doi.org/10.1093/biosci/bix125.

Rockström, Johan. 2015. "Bounding the Planetary Future: Why We Need a Great Transition." Great Transition Initiative. April. https://www.greattransition.org/publication/bounding-the-planetary-future-why-we-need-a-great-transition.

Rockström, Johan, et al. 2009. "Planetary Boundaries: Exploring the Safe Operating Space for Humanity." *Ecology and Society* 14(2): article 32. https://www.ecologyandsociety.org/vol14/iss2/art32/.

Rogelj, J., D. Shindell, K. Jiang, S. Fifita, et al. 2018: "Mitigation Pathways Compatible with 1.5°C in the Context of Sustainable Development. In: Global Warming of 1.5°C. An IPCC Special Report on the impacts of global warming of 1.5°C above pre-industrial levels and related global greenhouse gas emission pathways, in the context of strengthening the global response to the threat of climate change, sustainable development, and efforts to eradicate poverty" [Masson-Delmotte, V., P. Zhai, H.-O. Pörtner, D. Roberts, J. Skea, P. R. Shukla, A. Pirani, W. Moufouma-Okia, C. Péan, R. Pidcock, S. Connors, J. B. R. Matthews, Y. Chen, X. Zhou, M. I. Gomis, E. Lonnoy, T. Maycock, M. Tignor, and T. Waterfield (eds.)]. In Press.

Salk, Jonas. 1983. *Anatomy of Reality: Merging of Intuition and Reason*. New York, NY: Columbia University Press.

Salk, Jonas. 1992. "Are We Being Good Ancestors?" *World Affairs: The Journal of International Issues* 1(2):16–18.

Shove, Elizabeth and Gorden Walker. 2014. "What Is Energy For? Social Practice and Energy Demand." *Theory, Culture & Society* 31(5):41–58.

Stafford, Susan G. 2010. "Now Is the Time for Action: Transitions and Tipping Points in Complex Environmental Systems." *Environment* 52(1):40–45.

Steffen, Will, Paul J. Crutzen, and John R. McNeill. 2007. "The Anthropocene: Are Humans Now Overwhelming the Great Forces of Nature?" *Ambio* 36(8):613–621.

Steffen, Will, Wendy Broadgate, Lisa Deutsch, Owen Gaffney, and Cornelia Ludwig. 2015. "The Trajectory of the Anthropocene: The Great Acceleration." *The Anthropocene Review*, 1–18. https://doi.org/10.1177/2053019614564785.

Steffen, Will, Katherine Richardson, Johan Rockström, Sarah E. Cornell, et al. 2015. "Planetary Boundaries: Guiding Human Development on a Changing Planet." *Science* 347(6223). http://science.sciencemag.org/content/347/6223/1259855.

Strand, Ginger. 2008. "The Crying Indian." *Orion Magazine* 27(6). https://orionmagazine.org/article/the-crying-indian/.

Union of Concerned Scientists. 1992. "World Scientists' Warning to Humanity." https://www.ucsusa.org/resources/1992-world-scientists-warning-humanity.

United Nations Environment Programme. 2019. Emissions Gap Report 2019. November. UNEP, Nairobi.

Wallace-Wells, David. 2019. *The Uninhabitable Earth: Life After Warming*. New York, NY: Tim Duggan Books.

CHAPTER 9

The Only Constant

"Change is hard," a renowned chemist declared. She had just finished her presentation on the perils of plastic pollution and this was the start of her answer to a question about why we haven't yet addressed the awful and rapidly worsening problem. This conclusion puts her in good company. It's one of the go-to explanations for why we're not doing more about our urgent socio-environmental crises. In some ways, this unquestioned logic seems reasonable. Anyone who's ever tried to kick a habit or start a new one can attest to the challenge that presents. There's a lot of momentum keeping us on well-worn paths. Then again, people willingly embrace radical life-altering changes every day when they decide, for example, to change careers, get married, or have children. On a larger scale, consider the fact that the world's citizens went from using zero plastic drink bottles to consuming more than a million of them per minute in just a handful of decades. This is just one of countless examples of massive changes that have transpired in the recent past without most of us batting an eye or lifting a finger, so to speak. Maybe change isn't so hard after all.

In fact, one *could* see change as the easiest thing in the world. As ancient wisdom reminds us, it's the only real constant. At every scale, whether we're aware of it or not, change is *always* happening. Particles undulate. Energy flows. Cells are replaced or lost. Neural networks form

D. Kasper, *Beyond the Knowledge Crisis*,
Palgrave Studies on Norbert Elias,
https://doi.org/10.1007/978-3-030-48370-8_9

and re-form. Bodies mature, age, and die. Relationships develop, transform, and end. Societies change in ways big and small with the shifting composition of the people and patterns comprising them. Ecosystems undergo succession. A tilting planet spins and orbits around a star gradually burning up its hydrogen reserves in a solar system orbiting a galaxy which orbits other galaxies in a universe that is never standing still. In short, change is reality's default mode.

So, perhaps when people say "change is hard," they are referring more specifically to deliberate attempts to move in a direction that is counter to the momentum of dominant systems—be they in our minds, bodies, or societies. The resistance met here requires a great deal of energy to combat and is part of what makes some kinds of change more difficult. Reliance on tactics of direct intervention to force the desired outcome makes it more challenging still. Want to quit smoking or drinking? Just do it. Need to start exercising? Buckle down. Interested in eliminating plastic pollution? Stop buying and using the stuff. While determination and will power are helpful ingredients, they are rarely enough to make lasting personal change and are wholly inadequate for achieving the kinds of social change deemed necessary.

Recognizing these different realities of change (the ceaseless dynamism that characterizes all phenomena *and* the experience of friction in trying go against the grain) helps us understand why the changes involved in breaking bad habits or, say, reorganizing societies to support the health of Earth systems feel so hard, while also remembering that change is always already happening. Changes occurring at multiple levels and scales, even if we can't see them, create conditions which culminate in bigger, more observable changes. Understanding those conditions and working with *them*, rather than focusing primarily on the outcomes themselves, is the key to achieving meaningful and lasting change.

Whether we willingly undertake efforts to transform social systems in the ways deemed necessary to stay within safe planetary boundaries *or* permit growing instability in the Earth systems we depend on—which ultimately will impose its own uninvited effects—big changes are on the horizon. It's in our best interest to develop a more reliable understanding of how change works so that we can wield it to both guide its direction *and* gracefully navigate what comes. Presently, without the guidance of a scientifically-grounded theory ofsocial change, we're weak on both fronts. We need to do better; this chapter aims to show how we can.

It begins with a look at some of the main influences on how people make sense of social change and aspects of culture and social circumstance which help or hinder that. After reviewing prominent efforts to study social change, more scientifically, we find that confusion still surrounds it. This is in part due to some of the same obstacles encountered in attempts to explain the relationship between individual and society or society and environment and can be overcome by the same means: grounding people in the biophysical world and situating them in and among the overall pattern of interweaving socio-environmental processes. The chapter concludes with a discussion of how the socio-environmental synthesis framework functions as a theory of social change, with applications for explaining, anticipating, and guiding it.

Cultural Supports for Dealing with the Reality of Change

Day turns to night and then to day again. Years pass. Babies are born, children grow into adults, people age and eventually die. Human lives are defined by such transitions and by the gains and losses they inevitably bring. All cultures have rituals and rites of passage which mark time and remind us of change. Traditionally, cultural systems have, to varying degrees and in a variety of forms, provided aids for comprehending and coping with this basic fact of life.

One is the conventional wisdom they contain and disseminate. Respect for and practices related to the rhythms of tides, moons, and seasons, for instance, are ubiquitous in traditional cultures. In some, they are enshrined in classic writings, as with Ecclesiastes' well known instruction that "for everything there is a season" and that one's life, though it may appear solid, is like *hevel* (i.e., smoke or vapor). Efforts to concretize and hold it are in vain, like chasing the wind. Another such aid is language, which can facilitate thought and speech that more closely mirrors the reality of constant change, or doesn't. Like Potawatomi (mentioned in Chapter 4), Blackfoot is another example of several indigenous languages that superbly communicate dynamic relations. In contrast to noun-heavy English, Blackfoot "is about process, it's about actions" (Little Bear 2016). As such, it promotes a sense of the experience of living as "surfing the flux"—riding the waves of energy within an inter-relational whole—as opposed to proceeding along a linear path (Ross 1996:125). This kind of linguistic development has been central to enabling some cultures to remain harmonious with their geographic niche for thousands of years, notes systems scientist Peter Senge (2014). Third, and the integration of many such tools, are the highly developed philosophical systems which build an entire worldview around efforts to apprehend the reality of change. Buddhist thought is an example *par excellence*. Even with the extraordinary diversity apparent across different schools of thought in Buddhism, explicit recognition of impermanence is a shared foundation among them all. These kinds of cultural supports can help us comprehend the reality of constant flux—which the unfolding of time and the mathematics of contemporary quantum theory confirm—but which can be difficult to feel as true in our guts and minds in everyday life.

While it's encouraging to know that humans are capable of attaining and transmitting great wisdom about the reality of change, it's disheartening to see the extent to which supports for doing so have been lost, especially through the eradication of cultures that comes with the destruction of their landscapes and ecosystems. In my own context, many grow up today not only without reliable supports for understanding and accepting change, but must also contend with prominent dimensions of culture which work directly against it. An obsession with halting aging, or at least the appearance of it, the compulsion to fix moments in videos and photos, and the hiding away of death, for instance, are among the many cultural features that impede acceptance of the lived experience of change as a matter of course. This, combined with tremendous inconsistencies in people's scientific understanding of physical change processes and the obscurities imposed by the invisibility of far-flung figurations in mass society, poses serious challenges to our ability to make sense of social change. As a result, people hold wildly inconsistent ideas about how it works.

This is summed up well in one of my favorite *New Yorker* cartoons. In it, a person, dressed for the cold and standing outside a brick building, holds a picket sign that reads: "WE ARE BEING CONTROLLED BY THE RANDOM OUTCOMES OF A COMPLEX SYSTEM." In other words, we have a vague appreciation for the complexity of the world we live in, but in trying to locate the forces we perceive to be acting on our lives we instinctively place them somewhere beyond ourselves and characterize them variously as the results of either randomness or reasoned intent, or, as in this case, both.

Neither of these ideas is completely wrong, but neither do they get it quite right. There *are* plans, intentions, and randomness at play in the unfolding of social dynamics within their biophysical contexts, but these all interweave to:

> *give rise to changes and patterns that no individual person has planned or created. From this interdependence of people arises an order sui generis, an order more compelling and stronger than the will and reason of the individual people composing it.* It is this order of interweaving human impulses and strivings, this social order, which determines the course of historical change. (Elias 2000:366)

A theory of social change must explain "those long-term and unplanned, yet structured and directional trends in the development of social and personality structures that constitute the infrastructure of what is commonly called 'history'" (Elias 1997:355). Unfortunately, conventional thought habits make it difficult to grasp this, or at least to communicate about and act on it. Notable strides have been made in efforts to develop a more scientific study of social change, but there is a long way to go before the sociological sciences can explain social change in a way that is on par with explanations of changes in non-human nature.

Toward a Scientific Study of Social Change

Over time the natural sciences developed increasingly reliable ways to understand the processes that characterize non-human nature and thus to explain biophysical change. Along the way, a more scientific approach to studying social change was also pursued. This project, as we have seen, remains incomplete. Getting a glimpse of some of the most influential trends and countertrends in these efforts gives us an appreciation for the progress that was made and the work that remains to be done.

Shifting Views of Social Change in Western Thought

Most striking about early efforts to theorize social change is the about-face they represent in attitudes toward it. Elias outlines two key departures from traditional thought, especially apparent in the writings of Comte and Marx. For one thing, they broke free from the notion, firmly planted in Western philosophy of eternal reason and an unchanging mind, "supposedly shared by people in all historical periods and places" (Elias 1997:355). They showed instead, each in their own way, that human thinking about change itself changes in the course of time in accordance with historical circumstance. They also diverged from the standard view of social change as negative. For centuries, people saw human development as a decline, a fall from grace. Paradise and the golden age were thought to lay in the past which, at best, people could return to or foster a renaissance of. "From about the sixteenth century onwards there was a slow move away from the traditional placement of a high value on the past and being orientated to its authority, to a valorization of the present and the future" (Elias 1997:357). Marx and Comte exemplify how these tendencies gradually gave way to novel ideas about societies moving systematically through phases of change leading inevitably toward a better future.

The limited nature of knowledge at their disposal, Elias (1978) contends, made it possible for them to perceive clear trends in social development, filling in the gaps with inspired speculation influenced by the acute problems of their age. Regardless of the flaws in their thinking or differences in their respective social and political ideals, the recognition that current social relations represent only one moment in a long-term process leading from the past through the present and beyond it into the future was at the heart of their writings. As a result, both helped put the notion of social change at the forefront of human consciousness.

Interest in theories of law-like Progress, however, did not last. The discrepancy between the formulas of change proposed by early pioneers and the actual unfolding of modern history raised doubts about the idea of humanity's certain advancement. This was especially the case in the Western industrial societies of the twentieth century, where "the Enlightenment belief that by understanding the world we can improve the human condition" fell out of fashion, first among intellectuals and then the masses (Pinker 2018). It came to be considered, at best, naïve and, at worst, in service to racist, sexist, and hegemonic ends.

Yet faith in the future was not so easy to discard, resulting in certain palpable tensions. Alongside a deepening mood of doubt about the possibility of Progress, enthusiasm for progress—in the sense of humans being increasingly able to do once-only-dreamed-of-things—grew. Its advantages were accepted and eventually taken for granted, while its dangers were feared, their origins poorly understood. Commonsense accounts of the dangers humans posed for each other rarely transcended simplistic voluntaristic explanations, with threats and harms primarily attributed to the qualities and willful actions of nefarious others. A highly emotionally-involved view (including self-congratulation and self-preservation along with fear and loathing of the other) added to the difficulty of being able to read social processes more accurately. Over time, faith in the possibility of understanding and controlling those processes further waned. Yet, ironically, conscious concern and planning related to the improvement of the social order and living conditions continued to grow.

Although ideas of Progress had been discredited in the social sciences, the scientific study of social change remained a focal point. Indeed, Piotr Sztompka declares, it "is at the very core of sociology" (1993:xiii). What had been mostly lost, Elias observed, was the recognition that social change itself has an immanent order and structure. For all their faults, this was the great advance made by certain nineteenth-century thinkers. Having discarded this, later theorists (with some notable exceptions) predominantly portray human societies as generally harmonious and normally unchanging human constructions. "Social changes, often reified as 'social change,' appear in this theoretical approach as something additional, as disturbances in a social structure which would not change without disturbance" (Elias 1997:371).

Contemporary Treatments of Social Change

Even if one knows that change is the natural order of things, concepts that reinforce static pictures of social phenomena work against remembering and applying that awareness in the study of social change. Current assessments confirm that the impulse to translate processes into static conditions "still permeates sociology today—especially American sociology—in the form of *homo clausus*[1] and *Zustandsreduktion*[2]" (Górnicka et al. 2015). This is apparent in sociology survey textbooks, where the subject of "social change" typically appears in the last chapter, as if it were something additional, like change itself is presumed to be. Moreover, it is most often paired with discussions of social movements and collective behavior, underscoring the emphasis on voluntarism as what primarily propels change.

Writings dealing more exclusively with the subject of social change exhibit similar tendencies. Their overwhelming preoccupation with modernity, industrial capitalism, and modern globalization reflects the short-term, present-oriented perspective observed in the sociological sciences, in general. The propensity for listing "engines of change" (Weinstein 2010), "drivers of change" (Massey 2012), and "causes of social change" (Conley 2019; Harper and Leicht 2019) betrays a fixation on factors, as opposed to a more systems-based view including ongoing complex feedback processes. And, in spite of frequent efforts to frame large-scale social change within "the confluence of biography and history, the personal and the public, private efforts and the social milieu" (Massey 2012:2), the interface of those is never really clarified, keeping the individual/society divide firmly intact.

Often functioning, wholly or in part, as inventories of existing theories of social change, these works showcase the range of contributions to the theoretical repertoire of efforts aimed at understanding intensifying change processes (e.g., Chase-Dunn and Babones 2006; Harper and Leicht 2019; Noble 2000; Patterson 2018). As such, they represent significant accomplishments of description and rightly show that "there are many ways social change comes about" (Massey 2012:404). What they do not provide, however, is a reliable, shared, or actionable understanding of the basic pattern of social change within which one can make better sense of all those contributing factors. In the end, they mostly leave readers to "make their own choice or, better, to begin to work out their own arguments" about how social change occurs (Noble 2000:244).

It's reasonable to assume that these authors, like most social scientists, as Doug Massey points out, correctly "conceive of societal evolution as a complex of multicausal, multistranded social processes with nonlinear feedback loops that affect one another non-recursively over time" (2005:6). But without a foundational theory of social processes that accommodates complex systems and supports the development of our systems-thinking abilities, it is difficult to adequately communicate and act on that conception. Furthermore, it's impossible to devise such a theory when social change, like society itself, is viewed in isolation from biophysical reality. Social change is most often defined without any explicit integration into biophysical conditions, as, for example, the transformation of: "a culture over time" (Ferris and Stein 2016:467), "social institutions, political organizations, and cultural norms across time" (Conley 2019:780), or "social structure and cultural patterns through time" (Harper and Leicht 2019:5). This omission perpetuates the "human exemptionalism paradigm" observed by environmental sociology's founders.

There are some notable exceptions in efforts to produce "grand theories"—theories exhibiting "big vision about big processes that cut across big amounts of time" (Turner and Boyns 2001:353)—which promisingly feature human evolution, ecology, and geography as fundamental aspects of social change processes (Chase-Dunn and Lerro 2014; La

Piere 1965; Lenski 2005; Sanderson 1999). In contrast to standard treatments of social change, they: (1) do not leave assumptions about human nature unstated or unexamined, (2) look beyond the recent history of modernization, and (3) take the role of biophysical conditions in social life seriously (especially the phenotypical and neurological outcomes of evolution). There is much to appreciate in these efforts and their potential to influence thinking about social change. However, their tendency to include massive quantities of historical detail without an explicit encapsulation of any underlying theoretical framework makes it difficult to discern the theory being used or proposed, largely precluding subsequent applications of it. Lenski's (2005) ecological-evolutionary theory, represented in a neat visual model, constitutes an exception. Its utility, however, is somewhat limited by its attempt to identify a fixed set of significant variables (e.g., population, ideology, technology, social organization), the equating of "society" with nation-state, and its use of the concept of evolution as a parallel to socio-cultural change.[3]

The bottom line is that, despite a great deal of earnest scholarship, we still don't have a general theory of social change that is empirically grounded, widely recognized, and generally accepted as such. We have, rather, an agglomeration of theories—sometimes made explicit in teaching and research, more often not—which decidedly do not add up to a coherent understanding of social change.

Where Does This Leave Us?

All in all, this leaves us largely confused about how social change happens, ill-equipped to strategically bring about the kinds of changes being called for, and unprepared to skillfully navigate the changes that are coming. This collective confusion about the forces steering the trajectories of our lives and of history could be considered a fundamental social problem. Our mistaken assumptions thwart best-laid plans, result in misleading strategies or the belief that strategy itself is altogether pointless, and perpetuate the cycles of harmful behaviors we often feel stuck in. That's bad enough, but at a time when "whole-sale transformations of social and physical communities and dramatic changes in material life-styles" are necessary to stave off near-certain ecological and social catastrophe, the degree to which we grasp the mechanics and meanings of change is highly consequential (Rees 2019).

As we know, even if we choose not to make deliberate corrective efforts, we're still locked into massive changes. In Jim Kunstler's (2006) words, "the future is going to compel us to live differently, whether people like it or not, whether people want it that way or not. And this way of living is coming off the menu." Only with a firm grasp on the basics of social change processes can we hope to gracefully navigate the turbulence ahead.

Perpetuating Trained Incapacity

For these reasons, the scientific community long ago observed that "among the most general theoretical needs are improved ways to analyze social change," especially on longer time scales and across different levels of analysis and of spatial-temporal aggregation (Stern et al. 1992:41). Unfortunately, it appears to remain the case today, as it was then, that the "theory needed to understand global change does not yet exist" (Stern et al. 1992:198). This helps explain why even educational programs most strongly devoted to understanding and addressing socio-environmental problems offer students "blueprints of a sustainable future…but with little integrated, systematic sense of how to get from here to there" (Maniates 2013:268).

"Isn't it odd," someone commented in an Association for Environmental Studies and Sciences conference session, that all of "ESS [environmental studies and sciences] has a common preoccupation with systems and a natural science emphasis on how natural systems change, but we are oddly silent about how social systems change?" But even when they *are* talking about it, evidence shows that claims about social change are often contradictory and conveyed in implicit or otherwise uncritical ways. As a result, students report "that their education had not prepared them to critically analyze competing claims about social change" (Maniates and Princen 2015:216). Without a systematic way to compare and evaluate competing claims, they are left to "find their own way through the intellectual thicket," an approach that produces "cobbled-together notions of social change that manifest themselves as simplistic and counterproductive juxtapositions" (Maniates and Princen 2015:214).

So, not only are today's experts at a loss when it comes to questions of how to avert the disasters we're currently headed for, but our experts-in-training, for generations to come, will be similarly flummoxed by questions of social change and ill-equipped to navigate our uncertain futures. We can't really blame ESS programs for this. When even the literature specializing in social change concludes that, because the possibilities of theory are too diverse to be confined to any one framework, "we will always have to find our own way about amongst contradictory and irreconcilable perspectives," it is not surprising that educators emulate that view (Noble 2000:243). This has left people caught in a trap, Elias noted, with dystopian visions that reflect an:

> awakening disillusionment with the world as it is. So far they can only complain, as if someone owed them a better, more meaningful world. The traumatic shock, the mourning for lost illusions still blocks the realisation that no one can make this world better and more meaningful but they, human beings, themselves. (Elias 2009:273)

On the bright side, he jokes, citing estimates that the sun is only in the middle of its foreseeable lifespan, if humankind doesn't destroy itself first, "a future of 4000 million years should give humans the opportunity to muddle their way out of several blind alleys and to learn how to make their life together more pleasant, more meaningful and worthwhile" (1991:146). Might it be possible to integrate and put to use knowledge already at our disposal in ways that hasten this process? It is absolutely worth trying. "But," cautions sociologist Robert Brulle (2012), "we need to start from a realistic scientific basis."

In Need of a Different Kind of Sociological Theory

The main theoretical requirements for that basis, it's been observed, reflect the distinctive characteristics shared by human and environmental systems, including interdependencies, unanticipated consequences, nonlinearities between causes and effects, the potential for irreversible change, long time lags, and interactions of smaller-scale systems within the global system (Stern et al. 1992:174). That's a tall order. It's no wonder that a theoretical tool capable of accounting for all of that must begin at a high level of synthesis, a perspective from which we can glimpse the overall pattern of socio-environmental processes. The general insights attained here can then be applied to deeper study of particular phenomena and especially to specific questions about social change.

> "It is certainly not straightforward," Elias acknowledged, "to convert sociological theories which represent human societies, or even all humanity, as normally unchanging human constructions to a theory in which they become recognizable as endless processes" (1997:372).

But "without a fundamental shift of emphasis and direction, [sociology] can never form the basis for the social 'fund' of knowledge required to address the recurrent human crises out of which the subject itself was born" (Dunning and Hughes 2013:15). Straightforward or not, the need is great, and the demand for more effective ways to understand and guide social change is not going away. If sociologists aren't willing to take up the mantle others will, and have.

In one noteworthy example, a twenty-first-century team consisting of an evolutionary biologist and three psychologists uncannily echoes Elias' twentieth-century observations. Also concerned that "our ability to change our behavioral and cultural practices lags far behind our ability to manipulate the physical environment," they propose a science of intentional change (Wilson et al. 2014:395). They similarly attribute this lag to disarray in "human-related disciplines" and especially to the lack of a unifying framework within which to reconcile the supposed contradiction of genetic innateness and elaborate open-ended flexibility in humans. Adding to the challenge, they argue, is the fact that the concept of symbolic thought as a product of evolutionary processes and an important source of the human capacity for variation capable of producing new adaptations "is especially new for nearly all disciplines" (Wilson et al. 2014:415).

Despite differences between their project and the theory of social change proposed here, there is a shared conviction that coming to terms with the supposed paradox of human innateness and openness is imperative for transforming disciplinary paradigms in the ways needed to advance a shared reliable understanding of social change. Indeed, in taking biology seriously—but without treating it as "a solid bedrock that lies below cultural and social 'appearances'" (Meloni 2013:741) or privileging evolutionary science as the organizing theme (Smaldino and Waring 2014)—the socio-environmental synthesis framework does just that. After briefly considering the requirements of scientific sociological theory and the unique qualities of this one, we'll get into the details of how.

The Socio-Environmental Synthesis Framework as a General Theory of Social Change

"Facts are the world's data," wrote biologist Stephen Jay Gould. "Theories are structures of ideas that explain and interpret facts" (1981:35). It is in this sense that the proposed framework for socio-environmental studies is a theory. Deducing a handful of basic principles from among the infinite facts that could be observed about human social phenomena, the framework furnishes a structure for explaining and interpreting those facts. Briefly re-capping the premises articulated in Chapter 4, we recall that humans are biological and social organisms who are eminently changeable as a result of great neural malleability which both enables and necessitates social learning. Consequently, we exhibit a high degree of social interdependence, and therefore, human societies can be properly understood as dynamic patterns of interdependent social relations. These pillars of the socio-environmental synthesis framework help organize, explain, and interpret the world's data at the social level and, grounded in biophysical reality, are especially relevant to questions of interest in socio-environmental studies.

This framework also meets more specific criteria for scientific theory. It is "explicitly cumulative and integrating," with general principles that are "true across the widest range of application," and it "connects specialties together" (Collins 1986:1345). Additionally, genuinely theoretical formulations are "subject to falsification, explain the full range of phenomena under investigation in the most parsimonious fashion, order known (and predict unknown) facts, and demonstrate innovativeness and logical integration" (Michalski 2008:537). Somewhat less formally, Elias reminds us, theories resemble maps. "Like maps, theoretical models show the connections between events which are already known. Like maps of unknown regions, they show blank spaces where the connections are not yet known. Like maps, they can be shown by further investigation to be false, and they can be corrected…in contrast to maps, sociological models must be visualized in time as well as in space" (Elias 1978:160). This last point brings us back to the subject of social change, the focus of this framework. Relative to other social change theories, it has some unique qualities.

One is its universal applicability. Free from the narrow confines of a modern-, present-, or industrialized West-oriented perspective, one can profitably use the framework to understand social changes anywhere and anywhen social life among *Homo sapiens* takes place. Second, in situating social phenomena firmly in biophysical conditions, it is aligned with reality in ways that theories which either neglect the biophysical world or are based on dualistic pictures of our relationship "with" it are not. The next two qualities derive from the model's five-dimensional nature. Although time, the fourth dimension, can in some ways be considered an inherent feature of theories of social change, conventional static concepts make it difficult to convey, especially nonlinearly, as this model does. And, unique to the human social realm are the dynamic and shared systems of symbols and meanings that orient our perceptions and practices. This "fifth dimension" (Elias 1991:47), attention to which is often restricted to "micro" theories in conventional systems of thought, is integrated here in habitus. Finally, this theory of social change comes in a convenient and concise package, facilitating its conveyability and application—both highly unusual qualities among sociological theories. Let's now take a closer look at that the model representing both an organizing framework for socio-environmental studies and a theory of social change (see Fig. 9.1).

Over the past four chapters, we've considered the model's main components: biophysical conditions as the medium of human social life,

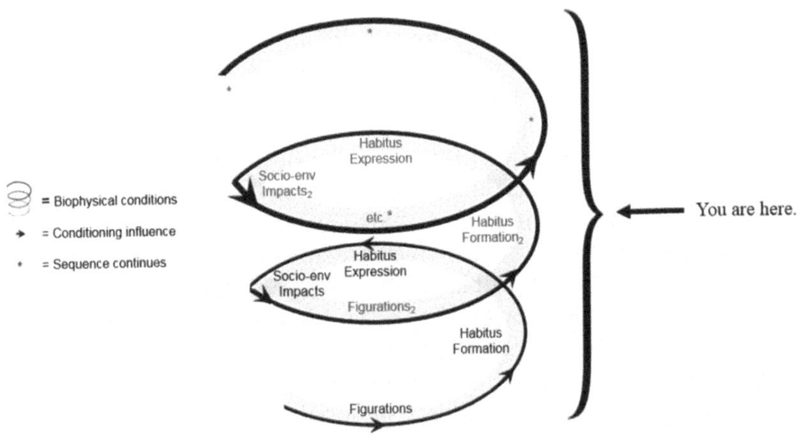

Fig. 9.1 Basic pattern of socio-environmental processes over time. This figure reminds us that every "moment"—comprised of and shaped by key interdependent dynamic components—emerges from what was there before and forms the basis of what comes next

interdependence via figurations as the human condition, the formation and expression of habitus as "second nature," and the inevitable impacts of human activities. Together, these form the basis of the general pattern of socio-environmental processes. On the one hand, we can see that these processes are happening concurrently in the unfolding of a given "moment" of time. At the same time, because their many underlying processes occur at different levels and temporal scales, we can also understand that there is a chronological order at work. The spiral shape, including "moments" and forward momentum, depicts the coinciding realities of apparent stability at a given time and the ongoing change also taking place.

We've contemplated the fact that change is a constant at every level of biophysical conditions, but it's important to note that not all such change can be considered *social* change. Nor are all socio-environmentally consequential impacts expressions of habitus. Random human and non-human events can also have impacts which change the course of history. So what is "social change?" Within the context of this theory, and adapting Elias' words slightly, social change can be defined as *the long-term and unplanned, yet structured and directional trends in the development of social and personality structures which embody, occur in, and are in relations of mutual influence with biophysical conditions* (see Elias 1997:355). Attention to individual personality structures and biophysical conditions in addition to social structures distinguishes this definition from standard ones. And in the context of the socio-environmental synthesis framework, the more expansive understanding of social change it confers generates valuable insights for efforts to explore and explain, anticipate, and guide social change.

Exploring and Explaining Social Change

Constructed around a small set of principles, rather than a list of variables deemed significant to a particular question in a specific historical context, this framework provides a reliable means of assembling the world's data and a systematic, albeit highly flexible, way to work with them to explore and make sense of social change. Because those principles—and the components of the theory they suggest—encompass phenomena at physical, biological, and social levels, the resulting theory facilitates the organization, integration, and sharing of knowledge across disciplines that is sorely needed but presently difficult to do in any systematic way.

Recalling the shelving analogy from Chapter 5, we might say that this framework gives us a place to "put" things, that is, the data, information, research questions, and findings relevant to complex socio-environmental problems of interest. In this way, it provides the coherence long sought in socio-environmental studies, where the greatest challenge has been figuring out how to intelligibly organize overwhelming quantities of knowledge from dissimilar fields and encourage meaningful collaboration across disciplines.

Beyond its benefits for organizing and synthesizing information, the socio-environmental synthesis framework provides the basis for doing what a genuinely sociological theory of change must do: facilitate efforts to explain general social change processes and specific trends observed. In contrast to standard depictions of social change as something that needs to be "added on" in order to account for disturbances in social structure, this framework naturally leads one through considerations of long-term social change processes. The components that make up individual "moments," which together comprise the trajectory portrayed in Fig. 9.2, conveniently translate into a series of interrelated questions about impacts, habitus, figuration, and biophysical conditions. Answers

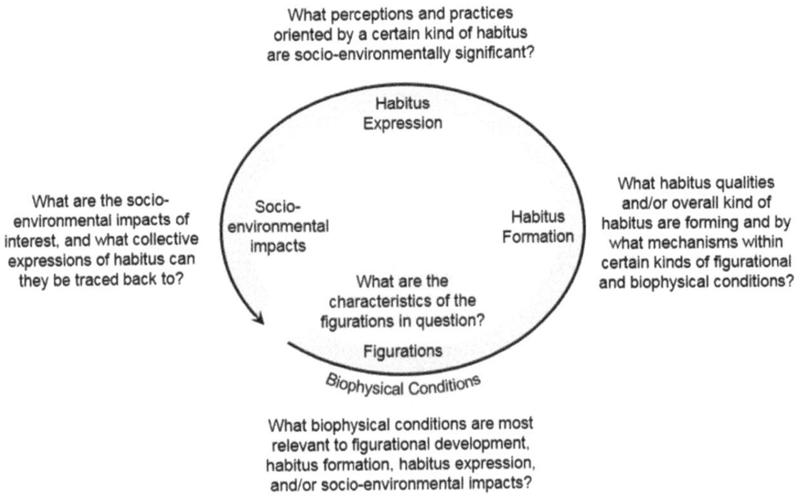

Fig. 9.2 A series of interrelated questions for informing efforts to explore and explain social change

to these questions at multiple "moments" of interest contribute to explanations of observed trends and long-term social change.

Taken separately or together, these general questions represent a systematic and iterative process for inquiry. Tailored to a specific topic, they highlight useful starting points and provide clear yet adaptable direction for researching questions about particular changes. For example, questions about particular impacts of the US food system and how to alter them, pursued in this framework, would prompt researchers to investigate the relevant practices and perceptions involved in production and consumption processes (e.g., standard farming practices, dietary preferences, norms related to eating), and under what conditions they exert what impacts. Following this further would lead to inquiry into the formation of underlying schemas, paradigms, and normalcy being expressed (e.g., humans vs. nature, reductionist view of ecological processes, prioritization of economic growth), the formation of those and overall habitus within certain figurational and biophysical conditions (e.g., the shape of market relations, information dissemination structures, dominant technologies), and ways that different figurational and biophysical conditions have supported the development of different sorts of habitus (e.g., comparisons of the above across different places and/or times).

Where to begin and which question or questions to take on will depend on the interests and capacities of the researchers. The limits of space permit only the most general allusions to possible uses; the main point is that this framework allows us to situate the questions, and the areas of study best suited to explore them, in and among one another in a way that was not possible before. Importantly, a more conscious sense of where a given question fits and how it relates to other work not only makes it easier to see opportunities for cooperation and to collaborate in developing explanations of social change, it lends itself to a systematic approach for anticipating it.

Anticipating Change

Does saying that this framework can help us anticipate change mean it allows us to make predictions? Well, yes, but not in the fortune-telling sense the word often connotes. "Every explanation makes possible predictions of one kind or another," Elias explains, but scientific prediction means different things at different levels of phenomena (1978:159). For one thing, the kinds of predictions one can make, and with what degree of certainty, vary with the degree of complexity of the phenomena under investigation. Technically, even the simplest physical phenomena can only ever be spoken of in probabilities. Uncertainty only grows as does the increasing complexity of the subject matter. Moreover, scientific prediction does not strictly pertain to future events (at least, not in the way we usually think of it). We cannot use the theory of evolution, for instance, to predict the future development of humankind, but we can use it "to predict that no human tooth could ever be found in a seam of coal—unless it had been put there by a miner!" (Elias 1978:159). If it were, the theory would need substantial correction and is, in this sense, falsifiable.

With regard to a *sociological* theory of change, we are reminded that figurations, formed by interdependent dynamic people, are flexible. Consequently, we know that the figuration at any later stage of development is only one of many possible transmutations of an earlier one and that, as changes in a particular figuration settle into place, the wide range of possible transformations narrows and the possibilities for what comes next are different than they otherwise might have been. While they don't function as prophecies, the kinds of predictions this theory makes possible allow us to make likely prognoses in two directions.

Looking backward to understand the development of a given condition, we can discern which earlier conditions (of figuration, habitus, etc.) were necessary for its formation. And, looking forward from the perspective of an earlier point, we can view some future condition as only one of a range of possibilities. In both cases, accumulated data about certain sociological conditions, combined with existing information about biophysical conditions, can enhance our ability to anticipate certain likely outcomes of certain combinations of conditions. Employing available information within the context of the socio-environmental synthesis framework prompts a novel and systematic series of questions relevant to understanding and anticipating how these circumstances could manifest in long-term social change processes (see Fig. 9.3).

Continuing with the food example, the following questions hint at how the above sequence of queries can be adapted to specific topics of interest. Beginning with a question about figurations and moving counter-clockwise through Fig. 9.3, we might ask, for instance, in what ways would a given climate scenario affect people's ability to meet food needs and wants and what kind of relational reconfigurations might be necessary or likely in order to meet those needs? (One example answer might be smaller, more localized food systems.) Under different

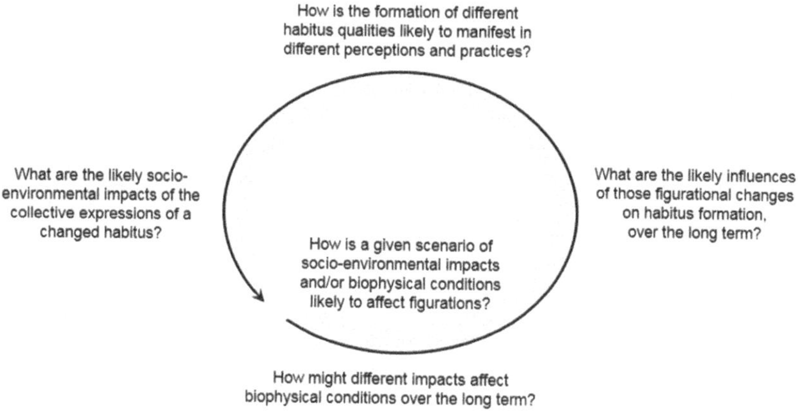

Fig. 9.3 A series of interrelated questions for informing efforts to anticipate social change

figurational conditions, will other conditioning influences on people's food-relevant "second nature" become more prominent? How might people's sense of "normal" develop differently (e.g., in smaller more local systems, would corporate advertising lose sway and processed food become a less typical part of people's diets?)? What kinds of activities, skills, and dispositions would be expressed via a different habitus developing in a different context (where food production is a normal facet of urban and residential landscapes, for example)? What values and beliefs would accompany a habitus formed within qualitatively and/or quantitatively different networks (prioritizing the ethical dimensions of food production, for instance)? How might different food-related practices and perceptions (such as the changes indicated above) increase or reduce environmental and social harms? How might they affect human health and quality of life? Over time, how might different impacts affect biophysical conditions and over what time frames (e.g., how might a decreased need for transport and storage affect greenhouse gas emissions? How would improved soil health influence carbon sequestration and water quality? How would diversifying crop production and re-foresting significant portions of agricultural land affect climate and local weather, and in what time frames?)?

Providing a checklist of primary considerations via the main components of the model, the proposed theory offers a systematic approach not only for explaining why conditions unfolded as they did, but also for anticipating likely future scenarios, given certain circumstances. Importantly, especially considering that the main impetus behind the current demand for a theory of social change is the desire for humanity to be able to steer away from the direction of catastrophic collapse and toward something better, the theory also offers novel insights into possibilities for more deliberately guiding social change.

Guiding Change

Besides enabling us to visualize socio-environmental processes over time without the strict dichotomization of individual and society and nature and culture, the alternative concepts at the core of this theory also render it useful for guiding efforts to more intentionally direct social change. In particular, its accommodation of human social interdependence via figuration and habitus clarifies key and mostly overlooked mechanisms of long-term social change. With these, we can more easily grasp the fact that changes in the organization of human relationships (i.e., figurations) go hand in hand with corresponding changes in people's conduct and personality structure (i.e., habitus). This insight, which Elias' work consistently demonstrates, is crucial to understanding the possibility of being able to more deliberately guide the direction of social change.

Charting the basic pattern of socio-environmental processes, this theoretical framework serves as a constant reminder that socio-environmental impacts (and their future implications) are generated by human activities oriented by a particular sort of habitus formed (and always forming) within a particular configuration of social relations, all taking place within a biophysical context and its particular conditions at a given time. Making meaningful and sustained changes to our impacts, this theory suggests, demands *changes in the conditions that give rise to them.* This points especially to the need to attend to figurations and the strategic modification of patterns of bonds of functional interdependence, reorganizing them in ways which support the desired changes in habitus, its expression, and impacts over time.

This is a shift away from the dominant approach of direct intervention toward a more indirect approach focusing on creating the conditions needed to achieve long-term goals. To demonstrate with an example from conservation biology, this is the difference between trying to address the problem of overpopulation in elephant reserves by directly manipulating their numbers through birth control and culling versus an indirect approach which seeks to alter the conditions within which the problem arose, rather than just attacking its symptoms. In one South African reserve, these methods involved creating ecological linkages among parks, removing fencing, reducing the number of watering holes, and generally trying to restore natural processes; and they worked (Belden 2013).

In the realm of socio-environmental problems, the direct approach has been dominated by the assumption that social change depends on the values and attitudes which drive behaviors that individuals rationally choose to adopt, what sociologist Elizabeth Shove calls the ABC model (signifying the standard emphasis on attitudes, behaviors, and choice). Influenced heavily by psychology and economics, this perspective "resonates with widely shared, commonsense ideas about media influence and individual agency" and it has defined relevant research, excluded or marginalized alternative perspectives, and shaped policies and strategies for action that in some ways have perpetuated or exacerbated problems and provide little help for comprehending or intervening in dynamic processes of social change (Shove 2010:1274). In contrast, the proposed theory seamlessly incorporates consideration of non-rational drivers into investigations of social change, the power and pervasiveness of which are increasingly recognized (Becker 1973; Quilley 2009; Seligman et al. 2008; Solomon et al. 2015). In this way, it provides a home for other alternative perspectives in socio-environmental studies. Most importantly, it fosters a more systemic and systematic examination of the social and biophysical conditions at the heart of human impacts. See Fig. 9.4 for a sequence of general questions to begin with in considering how to guide social change.

As above, some example questions related to concerns about the impacts of food systems are offered here to help illustrate the application of this framework for informing specific efforts to guide social change. Beginning with the desire to stem habitat and biodiversity loss (impacts), for instance, and moving counter-clockwise through Fig. 9.4, we might ask which farming practices are most consequential for species and habitat loss, and what specific changes in those would alter socio-environmental impacts in the ways needed or desired? From there, we might wonder what specific changes in farmers' and citizens' sense of normal would support and sustain those changed practices? What aspects of people's overall orientation toward food would support relevant changes in habits,

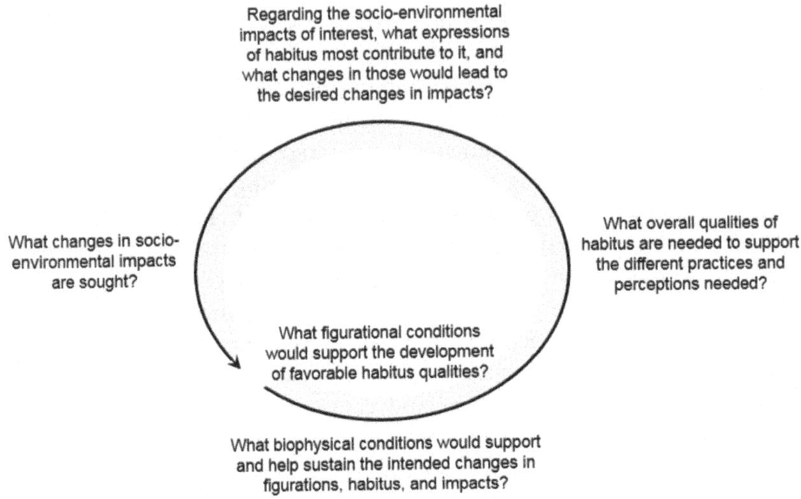

Fig. 9.4 A series of interrelated questions for informing efforts to guide social change

skills, worldviews, expectations, and tastes? With that in mind, we would inquire how the number of links and degrees of functional differentiation and integration in food supply chains could be modified to provide opportunities to develop and nurture the different desired skills, views, and habits related to food production and consumption? And are there rules and/or power differentials creating the conditions for and maintaining the status quo, in terms of "normal" practices? If so, how could they be altered to allow for the promotion of different ones? Next in the series are questions about biophysical conditions; for example, what landscape features would foster helpful changes in food-relevant figurations, habitus, and impacts? How could uses of technology be modified to do the same? And so on.

Although very general, the above examples help us begin to appreciate how the socio-environmental synthesis framework can help would-be changemakers see the range of possible entry points and where particular efforts from diverse disciplinary (and other) directions could best plug in. Then, considering the capacity available and the difficulty and timing of options, it becomes possible to make an informed choice about which efforts to support, advocate, or undertake—knowing where they fit in the bigger picture. It's important to note that the long-term perspective this framework affords does not preclude its ability to also inform immediate action to achieve near-term changes. To the contrary, it provides the context needed to be able to more responsibly undertake that emergency triage work, while at the same time advancing the long-term changes sought.

Caveats and Conclusions

Attitudes about our prospects for being able to use acquired knowledge to manage human progress continue to reflect a mix of enthusiasm, skepticism, fear, and a fair amount of disagreement about the details (as the numerous responses to Wilson et al. 2014 demonstrate). While the purpose of this book is not to advocate a science for managing social change, it does maintain that, in addition to explanatory and predictive power, a scientifically-grounded theory of social change offers valuable guidance for efforts to more intentionally direct social change. Soberly considering our potential capacity for guiding change, a few points warrant special emphasis here.

One is that, whatever concerns one might have about intentional efforts to make social change, the fact is that *whether* we contribute to shaping long-term trajectories of social change is not a choice we get to make. The effects may be subtle, but we are always participating in social change processes whether we know it or not. Among other ways, we do so by expressing our second nature within matrices of shared meanings and patterns of interdependent relations, and through our position on the receiving end of others' intentional efforts to effect certain kinds of change. Whether through policies created, carefully crafted campaigns conducted, edicts issued, protests waged, information and misinformation disseminated, emotions manipulated, and other tactics, our responses to these stimuli from our respective positions in relational power ratios figure into larger trends. Done with greater awareness, even these less direct forms of participation can be more impactful.

Secondly, like all self-organizing, nonlinear, complex systems with feedback loops, social systems are inherently unpredictable. In proposing a framework which can aid us in understanding and potentially guiding social change, this book makes no claims that social change processes are amenable to completely rational planning, management, or control. But neither can they be said to be completely irrational or random. Rather, as the theory shows, they are set and kept in motion by the dynamics of webs of relationships and the ways people are bound to live together within them. The reasonable conclusion drawn here, in agreement with Elias' assertion, is that, despite the impossibility of reason-based control, "it is by no means impossible that we can make out of it something more 'reasonable', something that functions better in terms of our needs and purposes" (Elias 1994:367). In other words, we *can* participate in and influence the direction of social change processes in ways that are informed by a more accurate picture of reality and sense of what's at stake with our behavior.

This is empowering, but we need to be clear on the fact that deliberate work of this type will take effort and time, the third point. It may turn out to be true in the end that the moral arc of the universe did bend toward justice, as Theodore Parker surmised and Martin Luther King, Jr. famously paraphrased. But here on Earth, the work of all that bending is up to flesh and blood people. Doing it successfully requires well-informed and prolonged effort. And even then, there are no guarantees that we'll achieve the changes we're after, or at least not in our lifetime. As difficult as it may be to accept, the social transformations needed to effect and sustain the changes our socio-environmental crises demand will almost certainly not happen as quickly as we'd like them to.

Elias' discussion of cycles of violence of war offers an analogy. Left alone, it may take a long time before the impetus of those processes exhausts itself, causing untold suffering along the way. The alternative—deliberate intervention to interrupt those cycles—requires a high level of detachment, self-restraint, patience, and a gradual toning down of negative feeling for each other and reconfiguring the pattern of relationships between them, from which the threat ultimately stems.

> None of these can be changed at short notice. Yet people often act as if that were the case: their emotional make-up induces them to expect and to demand that their political aims and ideals be realized in their own lifetime. (1987:xi)

Similarly, without intervention, the cycles of socio-environmental abuse and destruction large segments of humanity are deeply embroiled in would, in time, likely exhaust themselves, resulting in very different futures—visions of which are the stuff of contemporary dystopian fiction. Supremely preferable, however, is a deliberately charted alternative course seeking to maximize the health of Earth systems, ecosystems, and the human and non-human beings who depend on them. "Perhaps it is necessary to say explicitly," Elias interjects, "that work towards the fulfillment of many of these aims…is likely to require the patient work of many generations united in their uncompromising adherence to the paramount goal and ready for many compromises on detail" (1987:xi).

In any case, this deliberate work is only possible with greater capacities for detachment, self-restraint, and a transformation of attitudes toward and feelings about the other beings with whom we're deeply interconnected. All of these take time, but given the right conditions, the development of these capacities can be catalyzed and accelerated. For those still inclined to try given all this, the socio-environmental synthesis framework provides a crucial and previously unavailable tool. The next logical question is *what do we do with it?*

Notes

1. The image of a person as a closed-off, independent individual standing in opposition to others.
2. The literal translation is state reduction, the reducing of processes to static states.
3. Elias was careful to emphasize the difference between social development (denoting a neutral unfolding of social processes) and evolution (referring to biological processes occurring in a particular direction) in order to avoid "any implication of teleology, 'progress,' unilinearity, or inevitability, and stressing that social development, unlike biological evolution, can be reversed" (Mennell and Goudsblom 1998:12).

References

Becker, Ernest. 1973. *The Denial of Death*. New York: The Free Press.

Belden, Willow. 2013. "Abundance of Elephants Strains South African Game Reserves." All Things Considered, July 8. http://www.npr.org/templates/story/story.php?storyId=200104722.

Brulle, Robert J. 2012. Email Correspondence via American Sociological Association Environment and Technology Section Listserv, November 17.

Chase-Dunn, Christopher, and Salvatore J. Babones, eds. 2006. *Global Social Change: Historical and Comparative Perspectives*. Baltimore: The Johns Hopkins University Press.

Chase-Dunn, Christoper, and Bruce Lerro. 2014. *Social Change: Globalization from the Stone Age to the Present*. Boulder and London: Paradigm Publishers.

Collins, Randall. 1986. "Is 1980s Sociology in the Doldrums?" *American Journal of Sociology* 91(6):1336–1355.

Conley, Dalton: 2019. *You May Ask Yourself: An Introduction to Thinking Like a Sociologist*, 6th ed. New York and London: W. W. Norton.

Dunning, Eric, and Jason Hughes. 2013. *Norbert Elias and Modern Sociology: Knowledge, Interdependence, Power, Process*. London: Bloomsbury.

Elias, Norbert. 2009 [1981]. "What Is the Role of Scientific and Literary Utopias for the Future?" *Essays I On the Sociology of Knowledge and the Sciences, Collected Works* 14:269–287.

Elias, Norbert. 2000 [1939]. *The Civilizing Process*. Malden, MA: Blackwell.

Elias, Norbert. 1997. "Towards a Theory of Social Processes, a Translation," translated by Robert van Krieken and Eric Dunning. *British Journal of Sociology* 48(3):355–383.

Elias, Norbert. 1994 [1939]. *The Civilizing Process*, translated by Edmund Jephcott and edited by Eric Dunning, Johan Goudsblom, and Stephen Mennell. Malden, MA: Blackwell.

Elias, Norbert. 1991 [1989]. *The Symbol Theory*. London: Sage.

Elias, Norbert. 1987 [1983]. *Involvement and Detachment*, translated by Edmund Jephcott. Oxford: Basil Blackwell.

Elias, Norbert. 1978 [1970]. *What Is Sociology?* New York: Columbia University Press.

Ferris, Kerry, and Jill Stein. 2016. *The Real World: An Introduction to Sociology*, 5th ed. New York: W. W. Norton.

Górnicka, Barbara, Katie Liston, and Stephen Mennell. 2015. "Twenty-Five Years on: Norbert Elias's Intellectual Legacy 1990–2015." *Human Figurations* 4(3): http://hdl.handle.net/2027/spo.11217607.0004.303.

Gould, Stephen Jay. 1981. "Evolution as Fact and Theory." *Discover* May:34–37.

Harper, Charles L., and Kevin T. Leicht. 2019. *Exploring Social Change*, 7th ed. New York: Routledge.

Kunstler, James Howard. 2006. Appearance in "Radiant City," written and directed by Gary Burns and Jim Brown. Alliance Atlantis.

La Piere, Richard. 1965. *Social Change*. New York: McGraw-Hill Book Company.

Lenski, Gerhard. 2005. *Ecological-Evolutionary Theory: Principles and Applications*. Boulder and London: Paradigm Publishers.

Little Bear, Leroy. 2016. "Big Thinking and Re-Thinking: Blackfoot Metaphysics 'Waiting in the Wings.'" Big Thinking Conference, University of Calgary. https://www.youtube.com/watch?v=o_txPA8CiA4.

Maniates, Michael. 2013. "Teaching for Turbulence." Pp. 255–268 in *The State of the World 2013: Is Sustainability Still Possible?*, edited by Worldwatch Institute. Washington: Island Press.
Maniates, Michael, and Thomas Princen. 2015. "Fifteen Claims: Social Change and Power in Environmental Studies." *Journal of Environmental Studies and Sciences* 5: 213–217.
Massey, Douglas. 2005. *Strangers in a Strange Land*. New York and London: W.W. Norton.
Massey, Garth. 2012. *Ways of Social Change*. Los Angeles, CA: Sage.
Meloni, Maurizio. 2013. "Biology Without Biologism: Social Theory in a Postgenomic Age." *Sociology* 48(4):731–746.
Mennell, Stephen, and Johan Goudsblom. 1998. *On Civilization, Power, and Knowledge*. Chicago: University of Chicago Press.
Michalski, Joseph. 2008. "Scientific Discovery in Deep Social Space: Sociology Without Borders." *Canadian Journal of Sociology* 33(3): 521–553.
Noble, Trevor. 2000. *Social Theory and Social Change*. New York: St. Martin's Press.
Patterson, Thomas C. 2018. *Social Change Theories in Motion*. New York and London: Routledge.
Pinker, Steven. 2018. "How the Enlightenment Gave Us Peace, Prosperity, and Progress." Cato Institute Policy Report, April 9. https://www.cato.org/policy-report/marchapril-2018/how-enlightenment-gave-us-peace-prosperity-progress.
Quilley, Stephen. 2009. "The Land Ethic as an Ecological Civilizing Process: Aldo Leopold, Norbert Elias, and Environmental Philosophy." *Environmental Ethics* 31:115–134.
Rees, William. 2019. "Don't Call Me a Pessimist on Climate Change. I Am a Realist." *They Tyee*, November 11. https://thetyee.ca/Analysis/2019/11/11/Climate-Change-Realist-Face-Facts/.
Ross, Rupert. 1996. *Returning to the Teachings: Exploring Aboriginal Justice*. Toronto: Penguin.
Sanderson, Stephen. 1999. *Social Transformations: A General Theory of Historical Development*. New York and Oxford: Rowman & Littlefield.
Seligman, Adam B., Robert P. Weller, Michael J. Puett, and Bennett Simon. 2008. *Ritual and Its Consequences*. Oxford: Oxford University Press.
Senge, Peter. 2014. "Systems Thinking for a Better World." Aalto Systems Forum. Espoo, Finland. December 15.
Shove, Elizabeth. 2010. "Beyond the ABC: Climate Change Policy and Theories of Social Change." *Environment and Planning* 42:1273–1285.
Smaldino, Paul E., and Timothy M. Waring. 2014. "Let the Social Sciences Evolve." Commentary on "Evolving the Future: Toward a Science of Intentional Change," by Wilson et al. *Behavioral and Brain Sciences* 37(4):437.

Solomon, Sheldon, Jeff Greenberg, and Tom Pyszczynski. 2015. *The Work at the Core: On the Role of Death in Life*. London: Penguin Books.

Stern, Paul C., Oran R. Young, and Daniel Druckman. 1992. *Global Environmental Change: Understanding the Human Dimensions*. Washington, DC: National Academies Press.

Sztompka, Piotr. 1993. *The Sociology of Social Change*. Oxford and Cambridge: Blackwell.

Turner Jonathan H., and David E. Boyns. 2001. "The Return of Grand Theory." Pp. 353–378 in *Handbook of Sociological Theory*, edited by Jonathan H. Turner. Boston, MA: Springer.

Weinstein, Jay. 2010. *Social Change*, 3rd ed. Lanham, MD: Rowman and Littlefield.

Wilson, David Sloan, Steven C. Hayes, Anthony Biglan, and Dennis D. Embry. 2014. "Evolving the Future: Toward a Science of Intentional Change." *Behavioral and Brain Sciences* 37(4): 395–460.

Conclusion: Insights, Applications, and Possibilities

We've covered a lot of ground in the previous nine chapters. After reviewing the history of the knowledge crisis and the demand for synthesis, we looked at precedents for success, notable failures, and some of the major obstacles to the integration of knowledge sought, especially in socio-environmental studies. We saw how the subtly but significantly different way of envisioning phenomena hierarchically enabled us to more accurately situate humans in the world. With the resulting map in hand, from a satellite-like distance, we were able to discern key components of socio-environmental processes and detect the general pattern they form. From there, we zoomed in to get a closer look at each and how they, together, contribute to social change over time.

In one sense, I have not said anything new. I have merely repeated and assembled the conclusions of astute thinkers across the centuries, especially Norbert Elias and others who have similarly persisted in demanding more appropriate tools for thinking and speaking about human social life in terms of relational processes *in the world*. The insights their writings contain shed much-needed light on answers to some of today's most urgent questions, but it has been far too easy for them to get lost in the noise, casualties of the condition of having too much knowledge with too little organization. In this context, whether or not one encounters the insights needed is, as I once said about sociological knowledge, "largely the product of an unpredictable combination of personal inclinations and

D. Kasper, *Beyond the Knowledge Crisis*,
Palgrave Studies on Norbert Elias,
https://doi.org/10.1007/978-3-030-48370-8

dumb luck, not a recipe for a sound discipline," or interdiscipline like socio-environmental studies (Kasper 2011:121). This book is devoted to the premise that we can do better.

Ideas are a bit like water. If they are to be shared, especially in significant quantities and across some distance of space or time, a container is required. Without one, they are liable to slip through our fingers and the cracks. The proposed framework serves as a container for holding, conveying, and using ideas (and the questions, information, and knowledge underlying them) important to socio-environmental studies. In that sense, this project *does* offer something new. Analogous to the way novel properties emerge in organisms and societies when relationships among their constituent parts are reconfigured, so too a reconfiguration of relationships among the "parts" comprising the sprawling body of knowledge about socio-environmental processes generates a new kind of whole. The resulting synthesis, some have argued, holds the key to unlocking humans' ability to take a more intentional role in charting our course and shaping the future.

Reflecting on his original forecast, 15 years hence, Robert Heinlein elaborated on the role of the synthesist he then advocated. Its forerunners, he noted, were already at work in some places:

> they are all interdisciplinary people, generalists, not specialists—the new Renaissance Man. The very explosion of data which forced most scholars to specialize very narrowly created the necessity which evoked this new non-specialist. So far, this 'unspecialty' is in its infancy; its methodology is inchoate, the results are sometimes trivial, and no one knows how to train to become such a man. But the results are often spectacularly brilliant, too—this new man may yet save all of us [*sic*]. (1980:351)

Looser strictures on sci-fi authors allow him to make the sort of dramatic statements (about the salvation of humankind, for example) that scholarly writing prohibits. But we don't need to make or accept such audacious claims to be able to appreciate the underlying point: *an increased capacity to manage and make use of available knowledge cannot but be helpful in our attempts to gracefully navigate increasing turbulence and guide social change in ways that minimize harm and maximize health and happiness.* There are no guarantees, of course, but having the right tools for the job decidedly increases our chances of being able to more skillfully make our way through the world together. And, Elias (no date) affirms in a rare video recording that captures both his enthusiasm and his German accent:

> There is no greater task, no other task really, than, before destroying ourselves, to find out how we can arrange our lives in such a way that we do not constantly hurt ourselves, and also gain as much pleasurable excitement and satisfaction as we can.

"You will rightly say," he adds, "'how do we do it?'" The ending of this book is the beginning of an attempt to answer that question.

There is much work to be done, with the aid and expertise of many others, in discovering the potential value of a greater synthesis of knowledge for contributing to the transitions called for today—especially in terms of how we think, teach, research, and "arrange our lives." Below I take some initial first steps in discussing possible applications of the socio-environmental synthesis framework in education and research and invite readers to take things further in whatever directions suit them best to pilot, develop, refine, create, and share strategies and tools for advancing this crucial collective work. The book concludes with a discussion of some of the deeper psychological and cultural changes on which, some argue, the success of societal transformation will depend.

Education

Education can take many forms. With regard to integrating socio-environmental knowledge, understanding social change processes, and seeing our condition of dynamic interdependence more clearly, there are infinite possibilities for learning, in ways big and small, formal and informal. Here, I focus primarily on existing formal educational structures in higher education and how to adapt them in ways that serve those goals. The hope is that even a rough sketch of some potential next steps in a handful of academic arenas will prompt discussions of how these principles could be adapted within particular educational contexts and spur concrete action toward that end.

Knowledge Frameworks and Disciplinary Structures

By now, we're familiar with the bold yet common observation that "the structure of knowledge as presently and commonly perceived is being stressed and distorted by developments in contemporary society which, unless contained or accommodated, may destroy society itself" (Caldwell 1983:255). The creation of environmental studies was one attempt to accommodate growing knowledge and intensifying concerns about "environmental problems," as was the advent of environmental subspecialties in the social sciences and humanities. The difficulties experienced in trying to fit environmental concerns comfortably into any one discipline or in the overall structure of higher education, Caldwell argues, may be largely attributed to differences in perceptions of the environment and assessments of its significance as a focus of study. With the emerging understanding that what we call the environment constitutes a fundamental dimension of the reality in which humans exist and operate, however, the need to restructure the frameworks we use to understand human existence and activity to accommodate that becomes ever clearer.

While Caldwell stops short of specifying what a more workable knowledge structure should look like, leaving it as "a problem to be resolved" (1983:257), he emphasizes four points that are especially germane to considering potential modifications in academic structures: (1) all conceptual structures, including ways of perceiving and organizing knowledge, are artifacts of the mind and are thus mutable; (2) the organization of knowledge is not just a philosophical problem but, in our information-intense society, is everybody's concern; (3) the most sensible structure for organizing knowledge is hierarchical, where fields of knowledge of lesser complexity will be seen as underlying those of increasing complexity; and 4) unless we achieve a better means of integrating our expanding knowledge and relating its specialized components to one another in a hierarchical structure that more closely aligns with reality, the advancement and application of all knowledge may be detrimentally stalled. The socio-environmental synthesis framework enables us to build on these perceptive early observations. In sum, we can and should modify these structures and the three-tiered model of physical, biological, and social phenomena offers a simple but very useful basis for doing so. Below are three ways to apply it, starting with the easiest.

Categorizing Educational Offerings

Adjusting the way educational institutions categorize and present their offerings would require very little actual change in practice, while constituting a significant paradigmatic improvement. At present, there are essentially five major "buckets" into which academic programs are placed:

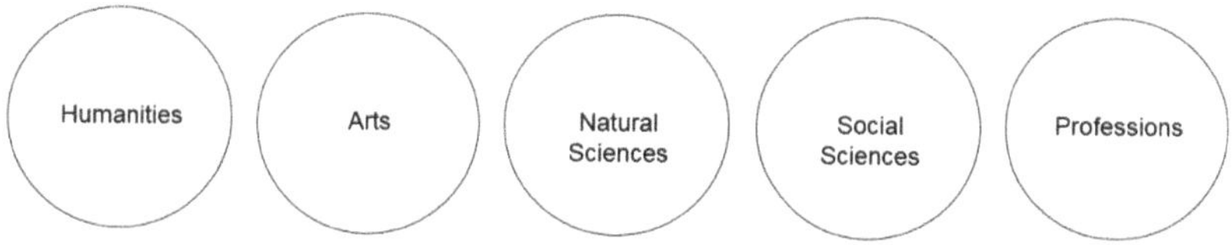

Fig. 1 Representation of unconnected programmatic categories in higher education

humanities, arts, natural sciences, social sciences, and professions. As with the social sciences discussed earlier, these categories are typically portrayed in an arbitrary side-by-side manner, with no indication as to how they might be connected (see Fig. 1). This could be seen as a fairly benign practice, except that it reinforces a fundamentally inaccurate sense of how the world's phenomena are related and represents a lost opportunity to revise that picture.

Alternatively, the simple hierarchical structure (with slightly more specific labels to accommodate additional distinctions) provides us with general categories of phenomena into which academic programs can more meaningfully fit, and in a way that is more in accord with reality (see Fig. 2). In it, physical and biological phenomena remain categories of their own, but social phenomena are divided into two areas. One deals explicitly with the study of social systems and processes and the other with "human pursuits"—the distinctively human activities we engage in (or do so in ways that other species don't), beyond those captured in the other three areas. This simple structure clearly signifies differences and junction points between these areas in a manner that conventional presentations do not, and is therefore conducive to cultivating a more accurate picture of how knowledge areas relate to one another within and across emergent levels. Like the small shifts described in the introduction, this subtle move makes a significant difference.

This basic logic can be applied in the categorization and representation of majors and minors, giving students a way to contextualize, at a glance, different courses of study within a bigger picture and begin to see connections among them. Figure 3, for example, shows where the majors and minors that existed at my institution at the time of this writing would fit.

There are prettier ways to depict this, no doubt, but this at least gives one a sense of the idea. The ability to recognize relevant relationships

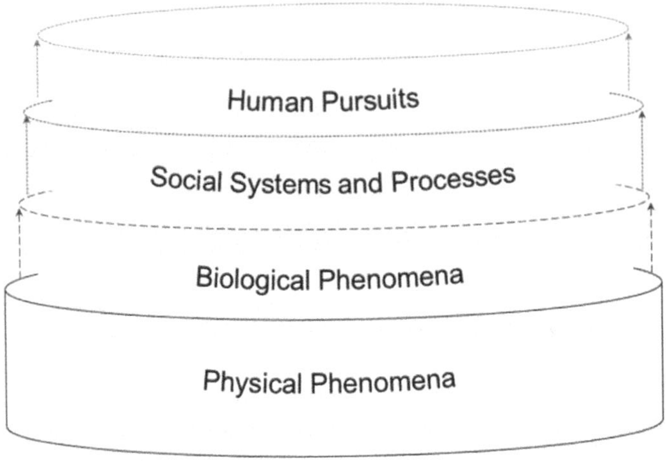

Fig. 2 General categories of phenomena into which academic programs can more meaningfully fit

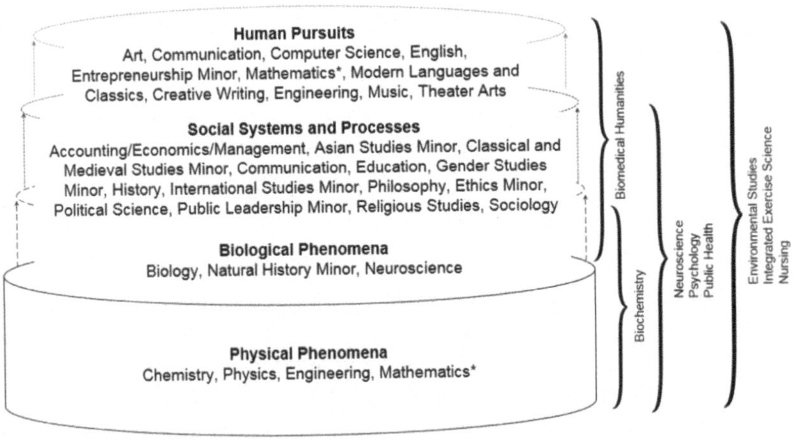

Fig. 3 Example of a college's majors and minors situated in the hierarchical framework. (*Due to conflicting opinions about whether mathematics belongs to the study of physical phenomena or represents the "the most extreme of the humanities" [Ashlock 2018], it is placed in both here.)

among seemingly disparate subject matter is a goal long advocated in socio-environmental studies and universally claimed by higher education professionals. Having a means of cultivating that ability in all students from their very first introduction to an institution, or to higher education in general, would represent real progress.

Taking this beyond mere presentation, this approach to situating areas of study (whether featured publicly or not) can also be useful in curricular planning and for informing a more sensible arrangement of academic programs. Instead of having to "create" or go to great lengths to explain interdisciplinarity in a curriculum, for instance, this approach makes clear that it's always already there because in actuality each higher level encompasses the ones "below" it, the study and interpretation of which is necessarily shaped by conditions in those "above." It also makes it possible to more clearly *show* the places where we work across multiple levels and thus span multiple disciplines. Finally, it aids teachers and administrators in more easily recognizing where seemingly disparate areas converge, supporting curricular design considerations and strategies for communicating about them.

Going bolder still, we could allow this structure to refine our vision of the social sciences to create a more appropriate and functional structure under the rubric of "sociological sciences" (as discussed in Chapter 3). As radical as that initially sounds, and as upsetting as it may be to some, even that is not really such a big deal. It need not significantly affect the names of majors and minors shown above or even departmental structures. Ideally, though, it *would* have meaningful effects on: how practitioners in those disciplines think about where they fit in the bigger picture, the fundamental paradigm with which they pursue their investigations, and how they portray that structure to others, especially the students whose worldviews they are helping shape via particular classes taught in their programs.

Individual Classes and Courses of Study

One of the greatest frustrations for students and new teacher-scholars is to have made the exciting discovery that solutions to major intellectual dilemmas exist, only to find out that they have been mostly ignored and have had little to no impact on the discipline in question. It's crucial that those in a position to represent an area of study are not only aware of its main challenges and unanswered questions, but that they address them directly when introducing students to a discipline. Providing this background can save students and subsequent generations of teachers and scholars the trouble of "having to discover these problems on their own—or worse, remain ignorant of them," and provides the context necessary for the disciplinary advancements needed (Kasper 2011:122). In addition to providing an integrative framework and theory of social change, this book serves other critical functions, especially in ESS and sociology. Below, I list and briefly describe some of the ways it can be used to meet learning goals in specific classes within these courses of study (see Table 1).

Research

As has been noted throughout this book, there is a great need for researchers, especially in socio-environmental studies, to be able to integrate, use, and engage in more interdisciplinary scholarship in order to comprehend and effectively deal with complex and inherently interdisciplinary problems. Particularly salient is "an urgent push to bridge the divide between the biophysical and the social sciences" (Brown et al. 2015). As we know, however, despite sincere efforts to clarify the human-environment interface and develop interdisciplinary research programs, the achievement of genuinely interdisciplinary research remains somewhat elusive (Roy et al. 2013; Wallace and Clark 2018). As coverage of this topic in the 2015 special issue of *Nature* affirmed, interdisciplinary research is harder to fund, do, review, publish, and garner institutional support for.[1]

These challenges, in large part, can be traced back to the difficulties funders, institutions, editors, reviewers, and even scholars themselves have conceptualizing the nature of, and empirical basis for, interdisciplinary research. In helping with *that*, the socio-environmental synthesis framework is inherently supportive of the development of more robust and integrative interdisciplinary research programs. Below, I mention some of the more specific ways it can be used in socio-environmental studies to illuminate connections among disciplines, highlight methodological developments needed, and point to alternative ways of organizing and presenting data.

Situating Disciplines and Specialties

Just as an organizational chart for a large corporation or government body allows us to see where different roles and the individuals who fill them fit

Table 1 Suggested uses of the socio-environmental synthesis framework and book chapters in specific classes

Class	*List of some general functions, needs, and goals of the class*	*Examples of suggested uses in service to those goals*
Intro to ESS[5]	✓ Introduce interdisciplinarity ✓ Master disciplinary concepts and understand how they relate and apply to concepts in different disciplines and to socio-environmental issues ✓ Clarify humans' relationship with nature ✓ Show long-term consequences of human actions and inaction on physical, biological, and social systems ✓ Demonstrate meaningful integration of disciplinary knowledge (especially across the natural and social sciences) in service to understanding and solving socio-environmental problems	✓ Chapter 1 is especially useful for providing the backstory of why interdisciplinarity is needed, some of the challenges associated with it, and why it's imperative to overcome them immediately ✓ Chapters 2 and 3 give further background on the challenges to interdisciplinarity and what's needed to overcome them, and are especially helpful for introducing certain natural science concepts to those unfamiliar and for those less familiar with the social sciences much-needed context for understanding what they are and the role they could play in socio-environmental studies ✓ Chapters 4, 5, 6, and 7, in particular, foster a sound, science-based understanding of humans' place in the world ✓ Chapters 8 and 9 clarify the consequences of human activities over time ✓ The socio-environmental synthesis framework (introduced in Chapter 4 and developed throughout the rest of the book) offers a means for integrating and using diverse knowledge

(continued)

Table 1 (continued)

Class	*List of some general functions, needs, and goals of the class*	*Examples of suggested uses in service to those goals*
Sociology Intro	Three particular classes are addressed below, though there are many other potential applications	
	✓ Define sociology ✓ Show how it relates to other disciplines ✓ Introduce core concepts and principles ✓ Explain why sociology is needed and what social functions it serves	✓ Chapter 3, or key parts of it, directly addresses questions of what sociology is, what it's for, and why we need it ✓ Chapter 3 suggests its proper relationship with other social sciences ✓ The basic principles in Chapter 4 form the basis for showing how it relates to other disciplines in a way that resolves long-standing riddles and questions ✓ Chapters 4, 6, 7, and 9 introduce and explain core sociological concepts and foundational principles ✓ Other parts of the book, namely Chapters 1, 2, 5, 8, and 9, clarify how environmental and other science themes integrate with sociology's core concepts and how sociology articulates with socio-environmental studies ✓ The socio-environmental synthesis framework offers a way to introduce a reliable general theory of human social processes from the beginning of their education in sociology[6]

Class	*List of some general functions, needs, and goals of the class*	*Examples of suggested uses in service to those goals*
Sociological Theory	✓ Convey the role of theory in building sociological knowledge ✓ Show how a theory organizes information to provide a coherent explanation of observed reality ✓ Provide a means for organizing numerous theories and theorists ✓ Demonstrate the complementary applications and mutually-influential nature of theory and method ✓ Instruct students in how to use a theory to organize, understand, and analyze data in relation to a specific question or problem	✓ This book clarifies what "sociological knowledge" is and how the theoretical framework assists us in building and conveying it ✓ The framework and the discussion of its components in Chapters 4–9 illustrate how diverse information derived from empirical observation can be integrated into a coherent whole to provide insights about social reality from a higher level of synthesis ✓ As an alternative to the unhelpful way sociological theories and theorists are usually organized (e.g., in terms of classical and contemporary, grouped by "perspective," or as a parade of individuals and their ideas), this framework provides a meaningful way to organize their works and/or the various concerns in them according to concern with: figurations (e.g., the consequences of relational structures in industrial capitalism, growing division of labor), habitus formation (e.g., the "I"/"me" dynamic, development of a Protestant ethic), habitus expression (e.g., spirit of capitalism, blasé attitude); socio-environmental impacts (e.g., suicide, land use), effects of these on figurational development and change over time (e.g., transition from mechanical to organic solidarity, globalization and world systems). These only begin to hint at the organizing potential of the framework, which would need to be worked out more fully elsewhere ✓ Relative to most existing sociological theories, the socio-environmental synthesis framework provides a firmer scientific basis and clearer procedures for applying it to organize information and ideas pertinent to investigating questions and problems of interest
Social Change[7]	✓ Explain basic mechanisms of social change processes ✓ Provide a way to evaluate and situate other social change theories ✓ Offer insights into how an understanding of social change could be effectively used to address problems, navigate turbulence, and guide the direction of change	✓ Chapters 5–9 address the basic mechanisms of social processes and their manifestation in short- and long-term change ✓ Grounded in empirical reality, the socio-environmental synthesis framework as theory of social change, discussed especially in Chapter 9, provides a basis for evaluating the soundness of other theories. It also gives us a way to place them in the framework, according to the specific levels and mechanisms with which they are concerned ✓ Chapter 9 and this chapter explicitly discuss ways to use the theory to shed light on possibilities for addressing problems, explaining and anticipating change, and guiding its direction more deliberately

in ways we cannot see from the ground, the synthesis framework gives researchers the means to situate themselves and their work within a much larger picture. Attuned to the main components apparent at its high level of synthesis and able to envision the layers of parts, processes, and mechanisms underlying them, we can more easily see where the expertise of different disciplines and specialties can best be applied. Figure 4 depicts, albeit in a general and only partial way, the novel possibility of being able to situate areas of study in relation to each other within the basic pattern of socio-environmental processes.

Examples of disciplines and specialties are placed in Fig. 4 according to their likely contributions to understanding subjects related to the main components in the model. Note, disciplines may fit in more than one area. Under habitus expression, for example, are those disciplines pertaining

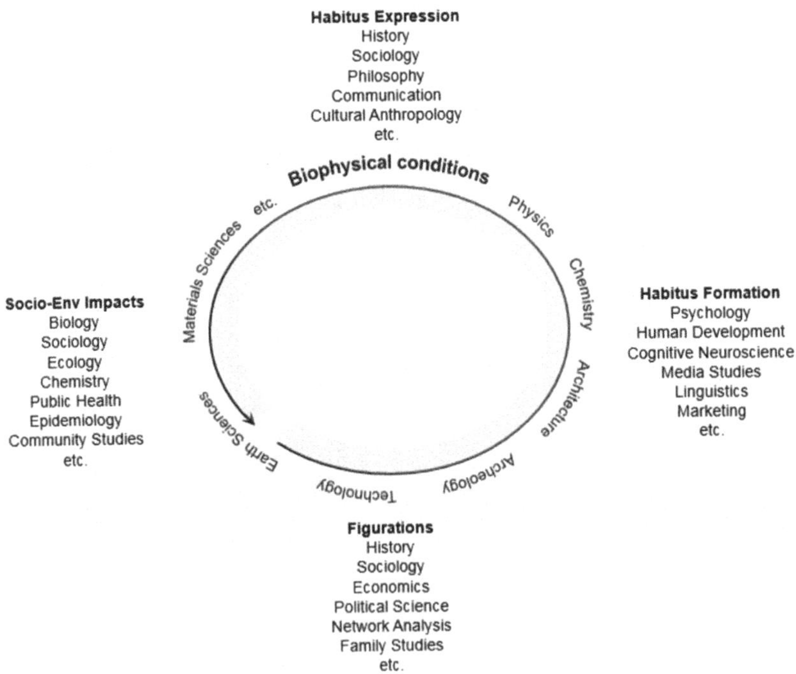

Fig. 4 Situating disciplines and specialties in relation to each other in the socio-environmental synthesis framework. Tentative and general example of possibilities

to study of the socially shared "normal" practices (e.g., activities, behaviors, habits, rituals) and perceptions (e.g., tastes, judgments, standards, thought patterns, worldviews) of people and their particular groups. Areas which investigate, at different levels, the mechanisms, processes, and contexts through which embodied schemas develop and are enacted contribute to our understanding of habitus formation. With regard to figurations, the various social sciences and other specialties focusing on particular categories of relations of functional interdependence (conventionally called institutions) and their structures, power ratios, and changes over time are positioned here. The same principle applies to research areas most relevant to socio-environmental impacts and biophysical conditions (of varying degrees of fixedness), the relevance of which is fairly intuitive in this framework.

This is just a start. Many more disciplines and specialties could be added to the lists above, but it at least begins to give an idea of the possibilities for more easily recognizing their contributions and connections. The intent here is not to imply a rigid formulation for these placements. Positions may shift with reference to different research questions and according to the unique needs of a project. The important point is that this synthesis framework enables us to systematically situate different focus areas relative to each other in ways we previously could not. This not only helps with seeing connections among existing research, but also with knowing where to look for information, identifying gaps and unanswered questions, and recognizing possibilities for collaboration.

Highlighting Methodological Needs

Clearly, there are already many well-developed methods for studying socio-environmental processes which can be put to use within the context of the proposed framework. There are also others needed which are either non-existent or in their infancy. In the sociological sciences especially, the move toward studying dynamic interdependence will require different kinds of methods, in addition to new theories and concepts. This subject is outside the range of my expertise, so I'll keep this section brief and general, but it's worth at least commenting on some of the theoretical-methodological opportunities to be seized by those in a position to do so.

Methods for assessing and mapping figurations and rigorously testing hypotheses about them, for example, stand in need of further development. While some have made contributions to thinking about process- and relational-oriented methodologies (Baur and Ernst 2011; Maguire 1988), continuation in this work would be welcome. I suspect there are also many others—especially among those with methodological interests related to network analysis, game models, complexity theory, big data, and more—who could help meet Elias' challenge of developing statistical aids suited to the task of investigating figurational processes (1978:132). The same could likely be said of available methodologies for studying the formation and expression of "second nature" and its socio-environmental impacts.

Paradigmatic transitions call for methodological transitions. The opportunity this framework presents to more directly ask and answer questions about which techniques are appropriate to the research task and which are not is especially welcome in the sociological sciences. Regardless of discipline, methods for making data, research findings, and complex socio-environmental systems more accessible through new means of organization and visualization are especially prized.

Visualizing and Organizing Data

The framework outlined in the previous five chapters is offered as part of a practical solution for getting us beyond our crisis of knowledge in socio-environmental studies. With its shift in perspective, and alternative concepts supporting that, it tames and transmutes the unruly dispersion of socio-environmental knowledge into a more comprehensible and useable form. It also serves as inspiration for the development of additional tools to support new ways of thinking and working. As Peter Senge (2014) has noted, in cultures where children are not supplied with tools, artifacts, and learning processes for cultivating their innate systems intelligence, it is vastly more difficult and, for some, impossible to use. Because this is the case in many cultures today, researchers and citizens alike may need some extra help understanding and imagining how to use the framework. I'll mention two possibilities here: a visual application to assist people in envisioning the many layers of parts and processes involved in socio-environmental phenomena at different scales and a structure for organizing research.

The visual tool I have in mind is a dynamic interface that gives users a "tour" of the processes represented in the socio-environmental synthesis framework. Beginning with a (more graphically sophisticated) picture of the overall pattern of interrelated processes, users can select components to explore and can then zoom in to move around and "up" and "down" at will through the underlying layers of parts and processes associated with each.

In a generic version of this, featuring key information and imagery of particular import in socio-environmental studies, "biophysical conditions" might include material related to basic Earth system conditions, ecosystem health, human biology, and the built environment. Influencing some of these over time are "socio-environmental impacts," which would guide one through information pertaining to, for example, some of the general ways human activities have affected atmospheric conditions, air, water, soil, ecosystems, plants, human health, communities, larger societies, and more at a range of scales from planetary to local. "Habitus expressions" would feature some of the most significant categories of activities, standard practices, habits, tastes, ways of perceiving, and more which are manifestations of "normal" in their respective contexts at scales ranging from individual to organization to mass society. "Habitus formation" would reveal the physiological and social mechanisms, processes, and contexts relevant to the development of schemas and embodied learning which come to feel second nature (e.g., activities in neurons and neural networks, language development, processes of interpreting and making meaning, and so on). Under "figurations" would be key qualities related to functional differentiation, levels of integration, and power ratios. Ideally, one would not only be able to explore these components at different spatial scales, but also through time to see what's changed and how quickly.[2]

This general idea could also be adapted to facilitate exploration of *particular* socio-environmental issues. Imagine, for instance, populating this structure with key information and images pertaining to water quality, climate change, food production, or high-consumption lifestyles. The beauty of it, in principle, is that it offers a way to envisage the whole (and the parts and processes comprising it) in ways we cannot directly observe in real life. Supporting the development of our ability to hold in mind and see connections among so many things at once, this interactive visual tool would benefit not only researchers from across a wide range of disciplines, but also those engaged in teaching and learning in socio-environmental studies. Realizing this vision would require the collaborative efforts of many. The hope in describing it here is that it may resonate with some who feel motivated and capable of helping make it happen.

The framework could also serve as a useful basis for the organization of existing socio-environmental research. Similar to the visual tool, the structure could be used to inform the location of data, research questions, studies, and/or key findings pertaining to a given subject according to their relevance to the framework's main components and the stuff underlying them.[3] This may be done formally in the context of large research databases, or informally as a means of organizing the literature relevant to a particular research project. Either way, it could conceivably facilitate the finding, and therefore also using, of existing and needed knowledge far more effectively than current practices, which at best rely on disciplinary categories and journal titles, and at worst on casting a line baited with keywords into the ocean of available scholarship and hoping to end up with a good catch.

Admittedly, it's easy to propose this organization in theory. In reality, its implementation would involve a lot of work—work ideally suited for synthesists! If the creation of a more intuitively navigable and theoretically coherent body of socio-environmental knowledge were deemed valuable, it would provide further motivation for creating and clarifying this role. And the fruits of their labors would likely have the added benefit of indirectly supporting deliberate efforts toward change.

Changemaking

As Chapter 9 showed, this framework provides much-needed guidance for exploring, explaining, and anticipating social change in a theoretically-grounded, disciplinarily-integrated, and systematic way. In so doing, it also furnishes those interested in *making* change with the essential tools they need to clarify long-term goals, anticipate eventualities, and identify a range of possible entry points. However one might participate in changemaking endeavors in service to addressing socio-environmental crises, the framework makes clear that significant and lasting changes in human impacts depend on changes in the activities underlying them, the habitus underlying those, and the figurational and biophysical contexts of both. The most novel and valuable insight here pertains to the crucial role of our connective patterns. Ultimately, re-conditioning ourselves and our activities in ways that engender and sustain the changes sought depends on the modification of our figurations. The key to becoming more effective changemakers, then, lies in cultivating the abilities to assess them, recognize trouble spots and opportunities, and identify the most promising places to cut, add, and rearrange in order to achieve the changes desired.

Although the guiding principles are constant, the details of their application are infinitely variable. For this reason, I don't try to offer a set of specific changemaking instructions here. Instead, I draw our attention to the more fundamental psychological and cultural changes we are being called upon to make at this time in human history: learning to see reality as it is, living within the limits that reality imposes, and accepting the responsibilities that await us. Only in achieving these, many argue, can we attain and sustain the societal transformations circumstances demand (e.g., in the energy sector, economies, climate policies, and other community resilience strategies). These two sides of the social change coin—changes in system design and operation and changes in how we think—are mutually supportive and reinforcing of one another. Having been likened to a kind of collective maturation, the various and cumulative metamorphoses these changes represent can be captured in a single phrase: *growing up*.

We Need to See Reality More Clearly As It Is

In his 1784 essay, "What is Enlightenment?" Immanuel Kant famously dared people to emerge from their self-imposed immaturity, which he defined as the inability to use one's understanding without the guidance of another. The enlightenment motto—*sapere aude* or dare to know, to use your own understanding—urged people to rely on their own reason rather than on pre-packaged truths passed down from authorities in their various forms. Two and a half centuries later, it is widely believed that "a social movement informed by the Enlightenment paradigm is likely to limit our collective capacity for social change" (O'Brien 2019). This means a lot of things (many of which have been addressed in earlier chapters), but in terms of the mandate to learn to see reality as it is, it means that we need to go *beyond* the conclusions we draw for ourselves based on how things *appear to be* from our limited perspective. If we were to express this idea in the form of a motto, *videre aude*—dare to see—might fit the bill.

While this sentiment has not yet manifest as a succinct rallying cry, it can be commonly found in socio-environmental literature in the form of ubiquitous demands for a new paradigm. Declaration of the need for paradigm shifts in how we think about economic growth, development, sustainability, environment, collective action, individualism, social change, and possible futures, for example, commonly appears in socio-environmental research and commentaries, often as a concluding thought and vague call to action. In contrast, this book *begins* by pointing out the need to see and think differently and then proceeds to offer tools for developing that capacity.

Of particular importance is the need to recognize the dynamic interdependence that characterizes ourselves and all of reality. We must come to recognize and viscerally know that, as Martin Luther King, Jr. eloquently put it in a 1967 Christmas sermon:

> all life is interrelated. We are all caught in an inescapable network of mutuality, tied together in a single garment of destiny. Whatever affects one directly, affects all indirectly…Before you finish eating breakfast in the morning, you've depended on more than half the world. This is the way our universe is structured, it is its interrelated quality. We aren't going to have peace on Earth until we recognize this basic fact of the interrelated structure of all reality.

King's point, though perhaps novel to many, is not new. It has long been the cornerstone of many traditional worldviews and has been corroborated by science time and again in countless ways. So, why have we, in the modern age, been so slow to catch on? And how can we catch up?

Attuning us to changes in social and personality structures over time, the synthesis framework helps us answer both questions. It points us toward empirical evidence of the rapid expansion and intensification of human interconnectedness in recent years, showing why, as Peter Senge (2014) observed, "our interdependence has grown and our awareness of the interdependence has declined." One modern manifestation of this is the way that a habitus shaped by traditional identities of more local groups lags behind the reality of the intertwined fates of peoples around the world. "People's whole outlook on life continues to be psychologically tied to yesterday's social reality, although today's and tomorrow's reality already differs greatly from yesterday's" (Elias 1995:35).

Just as a child's world grows larger with age, ability, and awareness—from the secure embrace of a loving caregiver, to the room a toddler is allowed to crawl or scoot around in, to the home and supervised outdoor spaces a new walker can explore, eventually expanding to include other people's homes, schools, neighborhoods and cities, travel destinations, and the whole world of places and people one has access to through books, television, movies, the Internet, and which ultimately everybody has to make their way in—so too we must allow our perception of reality to be continually adjusted as we observe and learn more, new, and different things. We must be able and willing to see what is revealed to us and allow it to change our view. In a context where our activities are more consequential than ever, the capacity to see the interdependent relations through which we wield those effects is especially crucial, as is the ability to exercise self-restraint in response to what those observations show us.

We Need to Recognize Limits and Learn to Live Within Them

Any physical entity with multiple inputs and outputs (that includes us) is surrounded by and interacts with layers of limits, making perpetual growth impossible (Meadows 2008). This seemingly unassailable logic has drawn fire from growth enthusiasts since it was first expressed. And in places like the United States, it has bumped up against a culture of extravagance where the doctrine of limitlessness, as Wendell Berry (2008) has called it, has "produced a sort of moral minimalism: the desire to be efficient at any cost, to be unencumbered by complexity… neighborliness, respect, reverence, responsibility, accountability, and self-subordination—this is the culture of which our present leaders and heroes are the spoiled children." This childish impulse to not want to be impeded by anything, to not want to be told "no," botanist and farmer Wes Jackson (2012) says, is behind the refusal to practice restraint. Failing to impose limits on ourselves, he observes, the system will impose them for us.

The illusion of limitlessness is starting to show cracks and the popular assumption that growth is the way to solve problems is gradually coming to be seen as the problem itself. In the intervening time, though, we have squandered precious opportunities to alter growth trends in ways that "establish a condition of ecological and economic stability that is sustainable far into the future" (Meadows et al. 1972:24). We have some catching up to do. Our job over the next century, Jackson (2012) argues, has got to be "devoted to acknowledging limits" and developing a culture of restraint, in which "progress" entails the denial of anything that chisels away at the ecosphere's capacity to nurture us.

Acknowledging that we are limited creatures in a limited world is neither condemnation nor resignation, but marks a return to our real condition from which we have temporarily been cut off. As earthly creatures, we naturally and inevitably live within limits, but as humans we get to choose how to respond to this necessity. Knowing who we really are enables us to do this with the confidence that our identity is not located "in the impulse of selfhood but in deliberately maintained connections" and the self-restraint they imply via values like stewardship, thrift, generosity, care, kindness, friendship, loyalty, and love (Berry 2008).

"Fully living within our finite ecosphere into the foreseeable future is or will be the most formidable challenge our species ever has or will ever face," says Jackson (2016). To inspire the kind of maturity this demands, we are in need of another kind of myth, increasing numbers of commentators agree. The narrative we need has been variously said to involve a new environmental ethics (Callicott 1987), a new ecological conscience (Quilley 2009), a new political narrative (Monbiot 2017a), a new civilization (Lent 2017), and a new Enlightenment (Hoffman 2019, Hoffman and Jennings 2018). Whatever it's called, to recover from our disease of limitlessness, says Berry (2008), we need to start with a different and much older premise: the naturalness and necessity of limits. "We must learn again to ask how we can make the most of what we are, what we have, what we have been given."

We Need to Accept Our Role and Responsibility as Co-Creators

When it comes to making long-term changes in what we're doing to the Earth and to each other, there's so much we don't know. "What we do know," author Bill Bryson observes, "is that there is only one planet to do it on and only one species capable of making a considered difference." As humans we are doubly lucky, we enjoy both the privilege of existence and the singular ability to appreciate it and, in many ways, to make it better. "It is a talent we have only barely begun to grasp" (2003:478). The question, then, is: What would we do, if we truly knew what we were doing?

For Ruben Nelson, executive director of Foresight Canada, this question is at the core of his life's work[4] and should be the driving question of our collective work together at this time in history. Our activities today are shaping the future; there's no getting around that. But what kinds of futures *could* we bring into being if we were equipped with a better sense of how that process works and of our role in it? Few are engaged in the work of trying to answer this question, and it's not on the formal agenda of any significant body. Though sustainability efforts were a promising start, some argue, significant amounts of the resources, energy, and attention devoted to them have been given over to trying to sustain modernity, when what we really need is "to learn to let it go as we out-grow it. The only way to grow is UP!" (Nelson 2019).

While many appreciate the need to get humanity off of its current trajectory, far fewer understand or can clearly articulate the nature of the work we face in correcting course. "The vision of a better world beckons," says Paul Raskin (2017) founder of the Great Transition Initiative, "but how do we get there?" is the question that dogs us. This book has discussed some of the academic shortcomings (i.e., fragmented knowledge and the lack of a theory of social change) that have gotten in the way of being able to offer sound answers to this question. In doing so, it has hinted at their relationship with a broader cultural sense of who and what we are, essential aspects of our "starting conditions." Others have looked more carefully at the role of cultural stories, especially the cumulative effects of being inculcated with visions of ourselves as separate, competitive, self-interested beings. The result, writer George Monbiot (2017b) argues, has been "a loss of belief in ourselves as a force for change, frustrating our potential to do what humans do best: to find common ground in confronting our predicaments, and to unite to overcome them."

As self-reflective meaning-making beings, it makes sense that our stories about ourselves and the world would have a profound impact on what we do and what we perceive ourselves to be capable of doing. This only underscores the need for a new story to support us through this collective coming of age, this wholly new experience of (potentially) transitioning into a more mature phase of agency as a species. Traditionally, cultures have provided individuals with stories, rituals, and rites to help carry them through the passage to adulthood, a transition which can at times feel scary and unappealing. We need supports like this on a much larger social scale. Part of the deep work of our time will be to articulate a new story of reality, one that reflects its (and our) dynamic and relational qualities and inspires us to more consciously wield that knowledge for the benefit of ourselves and others.

To say that our role, should we accept it, is to be co-creators of the future may sound fanciful or naïve to some, but it merely describes what's already taking place—only we are mostly oblivious to it and laboring under tragic errors. The humble suggestion here is that, in being able and willing to see more clearly, accept that which is revealed, and take responsibility for our actions, we can engage in transforming our civilization, ourselves, and our future with greater awareness and, preferably, in the ways being urgently called for.

For too long, though, the absence of a system for organizing interdisciplinary socio-environmental knowledge has held us back. Analogous to the difficulties of trying to locate needed information by sifting through heaps of spilled file cards, it has been a struggle to find, combine, and effectively use existing knowledge to address complex problems and steer social change. The socio-environmental synthesis framework developed in this book turns out to be more than just a vehicle for getting us beyond the knowledge crisis. It sheds much-needed light on social change processes and serves as a sound basis for the more accurate and compelling story people today have been charged with articulating.

Ideally, in the long run and through the cooperation of many, humans will come to understand and wield a greater degree of control over the dangers and benefits we pose for each other, just as advancements in science have enabled us to do with non-human nature. This future state will not be reached via inevitable or linear movement toward Progress. Nor can it be brought into being through force or ignorance or wishful thinking. In fact, overly strong emotionalinvolvements will only make this work more difficult, as Elias' reference to Poe's story of the maelstrom demonstrates (1987:45–118).

Allowing our wishes—built on hope, fear, and ignorance, as they are—to dominate our thinking, Elias adds in the video mentioned above, "prevents us from doing the first thing which has to be done, that is *to find out how our life together really works.*" This is the foundation on which any meaningful attempts at social transformation must be based and the main question this book is intended to help answer.

Notes

1. See volume 525, September 2015. https://www.nature.com/news/why-interdisciplinary-research-matters-1.18370.
2. Although this possibility requires further exploration, it's conceivable that the distance between layers in the spiral (or key points in a layer) could be used to indicate rates of change (i.e., the closer they are the slower the change, the farther apart the more rapid), thus conveying a great deal of information about complex dynamic phenomena at a glance.
3. See Kasper (2014) for a conceptual example of this applied to climate change research.

4. From his biography on http://foresightcanada.com/ruben-nelson-biography/, accessed March 21, 2020.
5. Similar uses apply to courses more generally intended to introduce interdisciplinary studies or more specific subjects like environmental humanities, social sciences, and methods in ESS.
6. One student, after reading part of the manuscript for this book, suggested that the socio-environmental synthesis framework would make an ideal outline for an introductory sociology text.
7. Social change courses could also appear in ESS curricula.

References

Ashlock, Daniel. 2018. "Math Is Not Science." Occupy Math blog, University of Guelph. https://occupymath.wordpress.com/2018/07/12/math-is-not-science/.

Baur, Nina, and Stefanie Ernst. 2011. "Towards a Process-Oriented Methodology: Modern Social Science Research Methods and Norbert Elias's Figurational Sociology." *Sociological Review* 59(s1):117–139.

Berry, Wendell. 2008. "Faustian Economics." *Harper's Magazine*, May. https://harpers.org/archive/2008/05/faustian-economics/.

Brown, Rebekah, Ana Deletic, and Tony H. F. Wong. 2015. "How to Catalyse Collaboration: Turn the Fraught Flirtation Between the Social and Biophysical Sciences into Fruitful Partnerships." *Nature* 525(315–317).

Caldwell, Lynton K. 1983. "Environmental Studies: Discipline or Metadiscipline?" *The Environmental Professional* 5:247–259.

Callicott, J. Baird. 1987. "The Conceptual Foundations of the Land Ethic." Pp. 186–217 in *Companion to a Sand County Almanac: Interpretive and Critical Essays*, edited by J. Baird Callicott. Madison: University of Wisconsin Press.

Elias, Norbert. 1995. "Technization and Civilization." *Theory, Culture & Society* 12(3):7–42.

Elias, Norbert. 1987 [1983]. *Involvement and Detachment*, translated by Edmund Jephcott. Oxford: Basil Blackwell.

Elias, Norbert. 1978 [1970]. *What Is Sociology?* New York: Columbia University Press.

Elias, Norbert. No date. Norbert Elias Foundation, videos. http://www.norbertelias foundation.nl/media/video/Elias2.m4v.

Heinlein, Robert. 1980. *The New Worlds of Robert Heinlein*. New York: Grosset and Dunlap.

Hoffman, Andrew. 2019. "Climate Change and Our Emerging Cultural Shift." *Behavioral Scientist*, September 30. https://behavioralscientist.org/climate-change-and-our-emerging-cultural-shift/.

Hoffman, Andrew, and P. Devereaux Jennings. 2018. *Re-engaging with Sustainability in the Anthropocene Era.* Cambridge, UK: Cambridge University Press.

Jackson, Wes. 2012. Interview with Aengus Anderson on "The Conversation: In Search of the New Normal." http://www.findtheconversation.com/episode-twenty-two-wes-jackson/.

Jackson, Wes. 2016. "Nature's Way: A Path to Ecological Agriculture." Interview, Great Transition Initiative, April. https://greattransition.org/publication/natures-way.

Kasper, Debbie V. S. 2014. "Codifying Figurational Theory and Mapping Common Ground in Sociology…and Beyond." 3(1): http://hdl.handle.net/2027/spo.11217607.0003.104.

Kasper, Debbie. 2011. "Finding Coherence in Sociology: (Finally) A Foundational Theory." Pp. 121–142 in *New Directions in Sociology*, edited by Ieva Zake and Michael DeCesare. Jefferson, NC and London: McFarland and Company, Inc., Publishers.

King, Martin Luther, Jr. 1967. "Christmas Sermon on Peace and Nonviolence, Massey Lecture #5." December 24. https://www.youtube.com/watch?v=1jeyIAH3bUI. 7:08-9:06.

Lent, Jeremy. 2017. *The Patterning Instinct: A Cultural History of Humanity's Search for Meaning.* New York: Prometheus Books.

Maguire, Joe. 1988. "Doing Figurational Sociology: Some Preliminary Observations on Methodological Issues and Sensitizing Concepts." *Leisure Studies* 7(2):187–193.

Meadows, Donella. H. 2008. *Thinking in Systems.* White River Junction, VT: Chelsea Green Publishing.

Meadows, Donella H., Dennis L. Meadows, Jørgen Randers, and William W. Behrens III. 1972. *The Limits to Growth.* New York, NY: Universe Books.

Monbiot, George. 2017a. *Out of the Wreckage: A New Politics for an Age of Crisis.* London: Verso.

Monbiot, George. 2017b. "How Do We Get Out of This Mess?" *The Guardian*, September 9. https://www.theguardian.com/books/2017/sep/09/george-monbiot-how-de-we-get-out-of-this-mess.

Nelson, Ruben. 2019. Email correspondence, SCORAI listserv. March 9.

O'Brien, Karen. 2019. "Activating Personal and Political Agency." Great Transition Initiative Forum: The Climate Movement: What's Next? June. https://greattransition.org/gti-forum/climate-movement-obrien

Quilley, Stephen. 2009. "The Land Ethic as an Ecological Civilizing Process: Aldo Leopold, Norbert Elias, and Environmental Philosophy." *Environmental Ethics* 31:115–134.

Raskin, Paul. 2017. "How Do We Get There? The Problem of Action." Great Transition Initiative Discussion Note, December. https://greattransition.org/publication/how-do-we-get-there.

Roy, Eric D., Anita T. Morzillo, Francisco Seijo, Sheila M. W. Reddy, Jeanine M. Rhemtulla, Jeffrey C. Milder, Tobias Kuemmerle, and Sherry L. Martin. 2013. "The Elusive Pursuit of Interdisciplinarity at the Human–Environment Interface." *BioScience* 63(9):745–753.

Senge, Peter. 2014. "Systems Thinking for a Better World." Aalto Systems Forum. Espoo, Finland. December 15.

Wallace, Richard, and Susan Clark. 2018. "Environmental Studies and Sciences in a Time of Chaos: Problems, Contexts, and Recommendations." *Journal of Environmental Studies and Sciences* 8(1):1–4.

Name Index

D. Kasper, *Beyond the Knowledge Crisis*,
Palgrave Studies on Norbert Elias,
https://doi.org/10.1007/978-3-030-48370-8

Subject Index

D. Kasper, *Beyond the Knowledge Crisis*, Palgrave Studies on Norbert Elias, https://doi.org/10.1007/978-3-030-48370-8

F

The manufacturer's authorised representative in the EU is Springer Nature Customer Service Centre GmbH, Europaplatz 3, 69115 Heidelberg, Germany. If you have any concerns regarding our products, please contact ProductSafety@springernature.com

Printed and bound by CPI Group (UK) Ltd, Croydon, CR0 4YY

22/09/2026

02221059-0002